S0-AUY-319

RELATIVITY

THE SPECIAL AND THE GENERAL THEORY

100TH ANNIVERSARY EDITION

PRINCETON UNIVERSITY PRESS
Princeton and Oxford

RELATIVITY

THE SPECIAL & THE GENERAL THEORY

100TH ANNIVERSARY EDITION

With commentaries and background material
by Hanoch Gutfreund and Jürgen Renn

ALBERT EINSTEIN

The translation by Robert W. Lawson is reprinted from *Relativity: The Special and the General Theory* (Crown, 1961 © The Hebrew University of Jerusalem)

Published by Princeton University Press, 41 William Street, Princeton, New Jersey 08540

In the United Kingdom: Princeton University Press, 6 Oxford Street, Woodstock, Oxfordshire OX20 1TW

press.princeton.edu

ISBN 978-0-691-16633-9

Library of Congress Control Number: 2015934627

British Library Cataloging-in-Publication Data is available

This book has been composed in Caslon 540 LT Std, Montserrat, and Relativ.

Printed on acid-free paper. ∞

Printed in the United States of America

10 9 8 7 6 5 4 3 2 1

CONTENTS

The great success in gravitation pleases me immensely. I am seriously contemplating writing a book in the near future on special and general relativity theory, although, as with all things that are not supported by a fervent wish, I am having difficulty getting started. But if I do not do so, the theory will not be understood, as simple though it basically is.

ALBERT EINSTEIN TO MICHEL BESSO, 3 JANUARY 1916

An etching of Albert Einstein by Hermann Struck, a Berlin artist who produced many such portraits of leading individuals of his time. This etching was included in several early foreign-language editions of the booklet.

INTRODUCTION

After submitting the final version of his general theory of relativity in November 1915, Einstein began to write a comprehensive summary of the theory for the scientific community. At that time he was already thinking about writing a popular book on relativity—both the special and the general—as he indicated in a letter to his close friend Michele Besso, quoted in the epigraph. Einstein completed the manuscript in December, and the booklet (as he referred to it) *Relativity: The Special and the General Theory (A Popular Account)* was published in German in the spring of 1917.

Einstein believed that the laws of nature could be formulated in a number of simple basic principles, and this quest for simplicity characterized his scientific activities. He also believed that it was his duty to explain these principles in simple terms to the general public and to convey the happiness and satisfaction that understanding them can generate. As Einstein stated in the short introduction to his booklet, he "spared himself no pains in his endeavour to present the main ideas in the simplest and most intelligible form," (p. 10) yet the book is not popular in the usual sense. It may be popular in its format, in its dialogue with the reader, in its examples from daily life, and in the lack of mathematical formulas, but it does not compromise on scientific rigor. The reader soon discovers that an intellectual effort is required to follow the flow of Einstein's thoughts and arguments.

In his brief introduction, Einstein also said that in the interest of clarity, he repeated himself frequently "without paying the slightest attention to the elegance of the presentation." To justify this approach, Einstein referred to the brilliant physicist Ludwig Boltzmann, "according to whom, matters of elegance ought to be left to the tailor and to the cobbler." Despite this assertion, the booklet is written with sophistication and elegance. The path from Newtonian mechanics to special relativity and then to general relativity—followed by its immediate consequences—emerges as an exciting intellectual odyssey. There is hardly any trace of Einstein's own bumpy road or of the difficulties he encountered on the way to this achievement.

However, Einstein was not happy with the result. In a letter to Besso he wrote: "The description has turned out quite wooden. In the future, I shall leave writing to someone else whose speech comes more easily than mine and whose body is more in order."[1] Jokingly, he later remarked that the description *Gemeinverstandlich* (generally understandable), on the cover of the booklet, should read *Gemeinunverstandlich* (generally not understandable).[2]

1 Einstein to Michele Besso, March 9, 1917, CPAE vol. 8, Doc 306, p. 293.

2 Pais, Abraham. *Subtle Is the Lord: The Science and the Life of Albert Einstein* (Oxford: Oxford University Press, 1982), p. 272.

Despite Einstein's self-criticism the booklet was a great success: 14 German editions appeared between 1917 and 1922, and a total of 15 were published in German during his lifetime. Strangely, though, the 15th edition, which appeared in 1954, was called the 16th instead. After the confirmation of the bending of light, the booklet was also published in many foreign languages.

In 1947, shortly after World War II and 30 years after publication of the first German edition, Einstein was approached by the publishing company Vieweg, which held the publication rights to the first edition, with a proposal to publish a new edition in German.[3] His response contained only two sentences in which he categorically rejected this proposal: "After the mass-murder of my Jewish brethren by the Germans, I do not wish any of my publications to be issued in Germany." Einstein's attitude softened over the years, and he approved the 1954 German edition, the last to appear in his lifetime.

Einstein's booklet is a unique document in the history of science writing. It is an attempt to enable the general reader with no background in physics to grasp and appreciate the grandeur of one of the most sophisticated intellectual achievements of the human mind and thereby to grant him or her, as Einstein puts it, "a few happy hours of suggestive thought." This goal may have been achieved in at least one case (see the letter from Walter Rathenau in the appended documents). Einstein appeals to the reader's intuition and does not assume that he or she has any previous knowledge of the subject matter. Familiar metaphors include train carriages and embankments. Einstein often poses a question to the reader—which he then answers

3 Einstein to Vieweg, March 25, 1947, unpublished, Vieweg Archive VIE: 18.

himself or in the name of the reader, taking both sides of a Platonic dialogue—which invites the reader to actively participate in the thought process.

Einstein maintained a close interest in the publication process and corresponded extensively with the publisher about the consecutive editions, translations into other languages, and granting rights to foreign publishers. He made stylistic and textual changes from one edition to the next, occasionally adding whole chapters and new appendixes. Thus, part III of the present version of the booklet—comprising sections 30–32, which deal with the universe as a whole—was added to the third edition in 1918. To the same edition Einstein also added the first two appendixes: "Simple Derivation of the Lorentz Transformation" (as a supplement to section 11) and "Minkowski's Four-Dimensional Space ('World')" (as a supplement to section 17).

The appendix "The Experimental Confirmation of the General Theory of Relativity" was written for the first English edition (1920) at the request of the translator. It discusses the three classical experimental tests of general relativity: the precession of the perihelion (the point of closest approach to the sun) of the planet Mercury, the deflection of light by the gravitational field, and the increased wavelength of spectral lines in a gravitational field (the *gravitational redshift*). This appendix was also included in the 10th German edition in 1920. The fourth appendix, "The Structure of Space according to the General Theory of Relativity," appeared first in the 14th English edition (1946) and was later included in the 1954 German edition. It is a supplement to section 32 and deals with the cosmological question of the nature of the universe. The term *space* here refers to the whole universe.

Of special character and significance is the long appendix "Relativity and the Problem of Space." It appeared first in the 15th English edition in 1954 and was then added to the 16th German edition in the same year. This appendix is very different from the others in that it reflects the development of Einstein's perception of the concept of space and is of a more philosophical character.

The text of Einstein's booklet, reprinted in the present volume, is the translation by Robert Lawson included in *The Collected Papers of Albert Einstein*, vol. 6, Doc. 42.[4]

The original text is accompanied by a "reading companion," which is a series of commentaries on the basic ideas, concepts, and methods that are the building blocks of the theory of relativity—the special and the general.

Following the reading companion is a chapter on the foreign-language editions, where we explore the history of and the stories behind the translations into foreign languages in the 1920s. We present this history in the context of the attitudes toward Einstein and his theories of relativity in the respective countries.

In line with Einstein's style and the nature of this book, our text contains only a few footnotes and references. Instead, we refer the reader to the most relevant sources and to a number of major works.

4 The authoritative edition of Einstein's papers is *The Collected Papers of Albert Einstein*, vols. 1–14 (Princeton, NJ: Princeton University Press, 1987–). This edition contains numerous invaluable introductions to the various aspects of Einstein's biography and work. The English translation volumes are referred to throughout. The published volumes of the *Collected Papers* are freely available online at einsteinpapers.press.princeton.edu. A substantial part of the Einstein Archives is made available online at www.alberteinstein.info by the Hebrew University of Jerusalem.

ACKNOWLEDGMENTS

We are grateful to Urs Schoepflin and Sabine Bertram from the Library of the Max Planck Institute for the History of Science, and to Roni Grosz, Chaya Becker, and Barbara Wolf from The Albert Einstein Archives at the Hebrew University of Jerusalem for helping us trace the different editions of the booklet and the relevant archival material. We are grateful to Diana Buchwald, the director of the Einstein Papers Project, for help and support in this project. We acknowledge with gratitude the help of Andrzej Trautman with the Polish edition, and the help of Danian Hu and Tsutomi Kaneko with the Chinese and Japanese editions, respectively.

Finally, we acknowledge with appreciation and gratitude the invaluable editorial assistance and professional support of Lindy Divarci.

RELATIVITY

THE SPECIAL AND THE GENERAL THEORY

100TH ANNIVERSARY EDITION

Einstein as a Missionary of Science

The dissemination of scientific ideas and discoveries to professional and intellectual circles outside a specific discipline and to the general public has always accompanied the development of science. Not only journalists and popular writers but also the scientists at the forefront of scientific activity have been involved in this endeavor. Thus Michael Faraday (1791–1867), who more than anyone else contributed to the empirical exploration of the laws of electromagnetism, delivered about one hundred public lectures, on topics including "domestic philosophy" and "a fire, a candle, a lamp, a chimney, a kettle and ashes." His lectures made the new experimental sciences, as well as the applied sciences of engineering and medicine, intelligible to the general public. James Clerk Maxwell (1831–1879), who cast the laws of electromagnetism into a coherent mathematical framework, wrote *Matter and Motion*, which is one of the most elegant elementary books on Newtonian dynamics. In modern times, many prominent scientists have written popular books on topics at the frontiers of contemporary physics. To name a few: *The First*

Three Minutes by Steven Weinberg, *The Quark and the Jaguar* by Murray Gell-Mann, and *A Brief History of Time* by Stephen Hawking.

A widespread image of Einstein is that of an isolated philosopher-scientist pondering the mysteries of the universe, far removed from everyday life. But this is a very misleading portrayal of his personality and his life. Einstein was a man of this world, collaborating and exchanging ideas with friends and institutions and acting as a politically engaged citizen. For four decades, from 1914 until his death, he articulated his views on every issue on the agenda of mankind in the first half of the twentieth century. In numerous articles, in correspondence with peers, and in public lectures he expressed his opinions on a variety of public, political, and moral issues, such as nationality and nationalism, war and peace, and human liberty and dignity. He also launched tireless attacks on any form of discrimination. Although Einstein expressed himself bluntly, was controversial, and was often considered simpleminded and naïve, his positions nevertheless made a significant impact.

Einstein's courageous public activities and his appreciation for the public understanding of science can be traced to a childhood episode he described in his *Autobiographical Notes*, written at the age of 67.[1] These notes constitute his scientific autobiography, but he included in these notes a single personal episode that, in his judgment, had a lasting influence on his personality and conduct in later years. His parents took into their home a medical student, Max Talmud, an orthodox Jew from Lithuania, to instruct little Albert in the principles of Judaism. However, at first

1 Einstein, Albert. *Autobiographical Notes: A Centennial Edition*, ed. Paul Arthur Schilpp (La Salle, IL: Open Court, 1992).

this young man proved much too successful, and to the dismay of his parents, Albert wanted them to keep a kosher home and observe other Jewish religious traditions. Einstein refers to this period as his "religious paradise of youth."[2] Fortunately, Mr. Talmud did more than provide instruction in Judaism—he also introduced Albert to Aaron Bernstein's *Popular Books on Natural Science*. Bernstein, a Jewish theologian, writer, and politician from Danzig, was a great popularizer of science. Albert was able to read there a chapter on a fantastic journey in space, which had a deep and lasting influence on his thinking. Einstein says that

> though the child of entirely irreligious (Jewish) parents—I came to a deep religiousness, which, however, reached an abrupt end at the age of twelve. Through the reading of popular scientific books I soon reached the conviction that much in the stories of the Bible could not be true. The consequence was a positively fanatic [orgy of] freethinking coupled with the impression that youth is intentionally being deceived by the state through lies; it was a crushing impression. Mistrust of every kind of authority grew out of this experience, a skeptical attitude toward the convictions that were alive in any specific social environment—an attitude that has never again left me.[3]

The consequence of this "mistrust of every kind of authority" was his independent approach, not restricted by conventional perceptions, which he applied to every issue he addressed—both within and outside science.

One aspect of Einstein's life and work that has not received the attention it deserves is his role as a missionary

2 *Autobiographical Notes*, p. 5.

3 *Autobiographical Notes*, pp. 3, 5.

of science, a popularizer, a communicator, an educator, and a moderator of science on the international stage. Einstein deliberately lent his name not only to political causes but also to the public dissemination of scientific knowledge on a worldwide level. Like few other scientists, he succeeded in conveying the results of his work to a broader public. Einstein not only published popular works and newspaper articles on his relativity theories but also held generally comprehensible lectures in publicly accessible venues, such as adult education institutions and planetariums. In February and March 1920, for example, he gave a series of 10 lectures on kinematics and equilibrium of bodies for the general public at the Adult Education College of Berlin. And in 1931 he famously lectured at the Marxist Workers' School on "What a Worker Needs to Know about the Theory of Relativity."[4] The playwright Berthold Brecht attended this lecture and was inspired by it in writing *The Physicist* (part of his famous anti-Nazi play *The Private Life of the Master Race*).[5]

Einstein's theory of relativity, the special and the general, presented a revolutionary worldview, with new insights on the concepts of space, time, and gravitation. The theory generated widespread interest and curiosity in broad intellectual circles and in the general public, creating an immediate need for authorized and accessible accounts of these ideas to dispel misunderstandings and to facilitate informed debates on the new ideas. Einstein felt compelled to respond to this need and thus wrote the present booklet, prior to which he had published three elementary essays—

4 Fölsing, Albrecht. *Albert Einstein: A Biography* (New York: Viking, 1997).

5 Brecht, Bertolt. *The Private Life of the Master Race* (New York: New Directions, 1944).

without the mathematical formalism—on the relativity principle and on the transition from the special to the general theory. This booklet, however, was the first comprehensive account after the final presentation of the theory.

Two years earlier, in 1912, Einstein had become a missionary of science in yet another sense: he had built a new bridge between physics and astronomy that was inspired by fundamental conceptual challenges of relativity and had far-reaching consequences for both physics and astronomy. To validate the predictions of relativity, it had been necessary to involve astronomers and to engage in a new form of collaboration between physics and astronomy. Ultimately, it became a new challenge to create and further develop an astrophysical community that until then had been practically nonexistent. Einstein made great efforts to motivate astronomers to check his predictions of general relativity, such as the gravitational bending of light and gravitational redshift. But his attempts were often met with indifference and even resistance. Only gradually did he succeed in rousing the interest of astronomers.

The turning point came in 1919, when the English eclipse expedition led by Sir Arthur Eddington confirmed Einstein's prediction of the gravitational bending of light. Thus, not long after the First World War, an English expedition had contributed to the success of the theory of a German-Swiss-Jewish scientist. In this way, science became a medium of international cooperation and Einstein its leading protagonist. Einstein became a celebrity practically overnight and proved well prepared to make judicious use of this prominence. From an early age, his thinking had been framed in internationalist and antimilitarist terms, and he felt that science should not be pursued as a narrow-minded, specialist enterprise. Einstein therefore took up the challenge of addressing the then-fledgling mass media and tried to explain aspects of his revolutionary theory to a wider public.

It is no exaggeration to claim that the public reception of general relativity and its creator contributed to a change in the societal status of modern physics on a global scale. First, quantum and, later, nuclear physics became an important driving force in enhancing the societal relevance of physics owing to their impact on a wide array of scientific fields and their real and potential applications. However, it was also the symbolic capital of Einstein's relativity revolution that helped establish physics as a leading paradigm of modernization, demonstrating that societal progress had become dependent on the progress of basic and not just applied science.

Einstein's personal contribution to this shift in perspective—as a cosmopolitan missionary of science during his travels—cannot be overestimated. He seems to have channeled some of the momentum of his youth that he drew from popular scientific culture to couple scientific and societal progress on a global scale. His journeys enhanced the processes of emancipation that were already underway in local scientific communities seeking a greater role for basic science within their societies. Of course, the local situations differed widely, from Spain to Japan, from Paris to Buenos Aires. A common feature of all the interactions between Einstein and the various scientific communities during his trips, however, was the significant increase in the awareness that basic science is a global endeavor of crucial relevance to all societies. In many cases, local-language editions of the present booklet preceded his visit and contributed to intellectual debates and to the public understanding of relativity.

Einstein's Booklet

Relativity

The Special
and the General Theory

by Albert Einstein

AUTHORIZED TRANSLATION BY
ROBERT W. LAWSON

Preface

The present book is intended, as far as possible, to give an exact insight into the theory of Relativity to those readers who, from a general scientific and philosophical point of view, are interested in the theory, but who are not conversant with the mathematical apparatus of theoretical physics. The work presumes a standard of education corresponding to that of a university matriculation examination, and, despite the shortness of the book, a fair amount of patience and force of will on the part of the reader. The author has spared himself no pains in his endeavour to present the main ideas in the simplest and most intelligible form, and on the whole, in the sequence and connection in which they actually originated. In the interest of clearness, it appeared to me inevitable that I should repeat myself frequently, without paying the slightest attention to the elegance of the presentation. I adhered scrupulously to the precept of that brilliant theoretical physicist L. Boltzmann, according to whom matters of elegance ought to be left to the tailor and to the cobbler. I make no pretence of having withheld from the reader diffculties which are inherent to the subject. On the other hand, I have purposely treated the empirical physical foundations of the theory in a "step-motherly" fashion, so that readers unfamiliar with physics may not feel like the wanderer who was unable to see the forest for trees. May the book bring some one a few happy hours of suggestive thought!

December 1916 A. EINSTEIN

Part I

The Special Theory of Relativity

ONE

Physical Meaning of Geometrical Propositions

In your schooldays most of you who read this book made acquaintance with the noble building of Euclid's geometry, and you remember—perhaps with more respect than love—the magnificent structure, on the lofty staircase of which you were chased about for uncounted hours by conscientious teachers. By reason of your past experience, you would certainly regard everyone with disdain who should pronounce even the most out-of-the-way proposition of this science to be untrue. But perhaps this feeling of proud certainty would leave you immediately if some one were to ask you: "What, then, do you mean by the assertion that these propositions are true?" Let us proceed to give this question a little consideration.

Geometry sets out from certain conceptions such as "plane," "point," and "straight line," with which we are able to associate more or less definite ideas, and from certain simple propositions (axioms) which, in virtue of these ideas, we are inclined to accept as "true." Then, on the basis of a logical process, the justification of which we feel ourselves compelled to admit, all remaining propositions are shown to follow from those axioms, *i.e.* they are proven. A proposition is then correct ("true") when it has been derived in the recognised manner from the axioms. The question of the "truth" of the individual geometrical propositions is thus reduced to one of the "truth" of the axioms. Now it has long been known that the last question is not only unanswerable by the methods of geometry, but that it is in itself entirely without meaning. We cannot ask whether it is true that only one straight line goes through two points. We can only say that Euclidean geometry deals with things called "straight lines," to each of which is ascribed the property of being uniquely determined by two points situated on it. The concept "true" does not tally with the assertions of pure geometry, because by the word "true" we are eventually in the habit of designating always the correspondence with a "real" object; geometry, however, is not concerned with the relation of the ideas involved in it to objects of experience, but only with the logical connection of these ideas among themselves.

It is not difficult to understand why, in spite of this, we feel constrained to call the propositions of geometry "true." Geometrical ideas correspond to more or less exact objects in nature, and these last are undoubtedly the exclusive cause of the genesis of those ideas. Geometry ought to refrain from such a course, in order to give to its structure the largest possible logical unity. The practice, for example, of seeing in a "distance" two marked positions on a practically rigid body is something which is lodged deeply in our habit of thought. We are accustomed further to regard three points as being situated on a straight line, if their apparent positions can be made to coincide for observation with one eye, under suitable choice of our place of observation.

If, in pursuance of our habit of thought, we now supplement the propositions of Euclidean geometry by the single proposition that two points on a practically rigid body always correspond to the same distance (line-interval), independently of any changes in position to which we may subject the body, the propositions of Euclidean geometry then resolve themselves into propositions on the possible relative position of practically rigid bodies.[1] Geometry which has been supplemented in this way is then to be treated as a branch of physics. We can now legitimately ask as to the "truth" of geometrical propositions interpreted in this way, since we are justified in asking whether these propositions are satisfied for those real things we have associated with the geometrical ideas. In less exact terms we can express this by saying that by the "truth" of a geometrical proposition in this sense we understand its validity for a construction with ruler and compasses.

Of course the conviction of the "truth" of geometrical propositions in this sense is founded exclusively on rather incomplete experience. For the present we shall assume the "truth" of the geometrical propositions, then at a later stage (in the general theory of relativity) we shall see that this "truth" is limited, and we shall consider the extent of its limitation.

[1] It follows that a natural object is associated also with a straight line. Three points *A*, *B* and *C* on a rigid body thus lie in a straight line when, the points *A* and *C* being given, *B* is chosen such that the sum of the distances *A B* and *B C* is as short as possible. This incomplete suggestion will suffice for our present purpose.

TWO

The System of Co-ordinates

On the basis of the physical interpretation of distance which has been indicated, we are also in a position to establish the distance between two points on a rigid body by means of measurements. For this purpose we require a "distance" (rod *S*) which is to be used once and for all, and which we employ as a standard measure. If, now, *A* and *B* are two points on a rigid body, we can construct the line joining them according to the rules of geometry; then, starting from *A*, we can mark off the distance *S* time after time until we reach *B*. The number of these operations required is the numerical measure of the distance *A B*. This is the basis of all measurement of length.[1]

Every description of the scene of an event or of the position of an object in space is based on the specification of the point on a rigid body (body of reference) with which that event or object coincides. This applies not only to scientific descrip-

[1] Here we have assumed that there is nothing left over, *i.e.* that the measurement gives a whole number. This difficulty is got over by the use of divided measuring-rods, the introduction of which does not demand any fundamentally new method.

tion, but also to everyday life. If I analyse the place specification "Trafalgar Square, London,"[1] I arrive at the following result. The earth is the rigid body to which the specification of place refers; "Trafalgar Square, London," is a well-defined point, to which a name has been assigned, and with which the event coincides in space.[2]

This primitive method of place specification deals only with places on the surface of rigid bodies, and is dependent on the existence of points on this surface which are distinguishable from each other. But we can free ourselves from both of these limitations without altering the nature of our specification of position. If, for instance, a cloud is hovering over Trafalgar Square, then we can determine its position relative to the surface of the earth by erecting a pole perpendicularly on the Square, so that it reaches the cloud. The length of the pole measured with the standard measuring-rod, combined with the specification of the position of the foot of the pole, supplies us with a complete place specification. On the basis of this illustration, we are able to see the manner in which a refinement of the conception of position has been developed.

(a) We imagine the rigid body, to which the place specification is referred, supplemented in such a manner that the object whose position we require is reached by the completed rigid body.

[1] I have chosen this as being more familiar to the English reader than the "Potsdamer Platz, Berlin," which is referred to in the original. (R. W. L.)

[2] It is not necessary here to investigate further the significance of the expression "coincidence in space." This conception is sufficiently obvious to ensure that differences of opinion are scarcely likely to arise as to its applicability in practice.

(b) In locating the position of the object, we make use of a number (here the length of the pole measured with the measuring-rod) instead of designated points of reference.

(c) We speak of the height of the cloud even when the pole which reaches the cloud has not been erected. By means of optical observations of the cloud from different positions on the ground, and taking into account the properties of the propagation of light, we determine the length of the pole we should have required in order to reach the cloud.

From this consideration we see that it will be advantageous if, in the description of position, it should be possible by means of numerical measures to make ourselves independent of the existence of marked positions (possessing names) on the rigid body of reference. In the physics of measurement this is attained by the application of the Cartesian system of co-ordinates.

This consists of three plane surfaces perpendicular to each other and rigidly attached to a rigid body. Referred to a system of co-ordinates, the scene of any event will be determined (for the main part) by the specification of the lengths of the three perpendiculars or co-ordinates (x, y, z) which can be dropped from the scene of the event to those three plane surfaces. The lengths of these three perpendiculars can be determined by a series of manipulations with rigid measuring-rods performed according to the rules and methods laid down by Euclidean geometry.

In practice, the rigid surfaces which constitute the system of co-ordinates are generally not available; furthermore, the magnitudes of the co-ordinates are not actually determined by constructions with rigid rods, but by indirect means. If the results of physics and astronomy are to maintain their clearness, the physical meaning of specifications of position must always be sought in accordance with the above considerations.[1]

We thus obtain the following result: Every description of events in space involves the use of a rigid body to which such events have to be referred. The resulting relationship takes for granted that the laws of Euclidean geometry hold for "distances," the "distance" being represented physically by means of the convention of two marks on a rigid body.

[1] A refinement and modification of these views does not become necessary until we come to deal with the general theory of relativity, treated in the second part of this book.

THREE

Space and Time in Classical Mechanics

The purpose of mechanics is to describe how bodies change their position in space with "time." I should load my conscience with grave sins against the sacred spirit of lucidity were I to formulate the aims of mechanics in this way, without serious reflection and detailed explanations. Let us proceed to disclose these sins.

It is not clear what is to be understood here by "position" and "space." I stand at the window of a railway carriage which is travelling uniformly, and drop a stone on the embankment, without throwing it. Then, disregarding the influence of the air resistance, I see the stone descend in a straight line. A pedestrian who observes the misdeed from the footpath notices that the stone falls to earth in a parabolic curve. I now ask: Do the "positions" traversed by the stone lie "in reality" on a straight line or on a parabola? Moreover, what is meant here by motion "in space"? From the considerations of the previous section the answer is self-evident. In the first place we entirely shun the vague word "space," of which, we must honestly acknowledge, we cannot form the slightest conception, and we replace it by "motion relative to a practically rigid body of reference." The positions relative to the body of reference (railway carriage or embankment) have already been defined in detail in the preceding section. If instead of "body of reference" we insert "system of co-ordinates," which is a useful idea for mathematical description, we are in a position to say: The stone traverses a straight line relative to a system of co-ordinates rigidly attached to the carriage, but relative to a system of co-ordinates rigidly attached to the ground (embankment) it describes a parabola. With the aid of this example it is clearly seen that there is no such thing as an independently existing trajectory (lit. "path-curve"[1]), but only a trajectory relative to a particular body of reference.

In order to have a *complete* description of the motion, we must specify how the body alters its position *with time*; *i.e.* for every point on the trajectory it must be stated at what time the body is situated there. These data must be supplemented by such a definition of time that, in virtue of this definition, these time-values can be regarded essentially as magnitudes (results of measurements) capable of observation. If we take our stand on the ground of classical mechanics, we can satisfy this requirement for our illustration in the following manner. We imagine two clocks of identical construction; the man at the railway-carriage window is holding one of them, and the man on the footpath the other. Each of the observers determines

[1] That is, a curve along which the body moves.

the position on his own reference-body occupied by the stone at each tick of the clock he is holding in his hand. In this connection we have not taken account of the inaccuracy involved by the finiteness of the velocity of propagation of light. With this and with a second difficulty prevailing here we shall have to deal in detail later.

FOUR

The Galileian System of Co-ordinates

As is well known, the fundamental law of the mechanics of Galilei-Newton, which is known as the *law of inertia*, can be stated thus: A body removed sufficiently far from other bodies continues in a state of rest or of uniform motion in a straight line. This law not only says something about the motion of the bodies, but it also indicates the reference-bodies or systems of co-ordinates, permissible in mechanics, which can be used in mechanical description. The visible fixed stars are bodies for which the law of inertia certainly holds to a high degree of approximation. Now if we use a system of co-ordinates which is rigidly attached to the earth, then, relative to this system, every fixed star describes a circle of immense radius in the course of an astronomical day, a result which is opposed to the statement of the law of inertia. So that if we adhere to this law we must refer these motions only to systems of co-ordinates relative to which the fixed stars do not move in a circle. A system of co-ordinates of which the state of motion is such that the law of inertia holds relative to it is called a "Galileian system of co-ordinates." The laws of the mechanics of Galilei-Newton can be regarded as valid only for a Galileian system of co-ordinates.

FIVE

The Principle of Relativity (in the Restricted Sense)

In order to attain the greatest possible clearness, let us return to our example of the railway carriage supposed to be travelling uniformly. We call its motion a uniform translation ("uniform" because it is of constant velocity and direction, "translation" because although the carriage changes its position relative to the embankment yet it does not rotate in so doing). Let us imagine a raven flying through the air in such a manner that its motion, as observed from the embankment, is uniform and in a straight line. If we were to observe the flying raven from the moving railway carriage, we should find that the motion of the raven would be one of different velocity and direction, but that it would still be uniform and in a straight line. Expressed in an abstract manner we may say: If a mass m is moving uniformly in a straight line with respect to a co-ordinate system K, then it will also be moving uniformly and in a straight line relative to a second co-ordinate system K', provided that the latter is executing a uniform translatory motion with respect to K. In accordance with the discussion contained in the preceding section, it follows that:

If K is a Galileian co-ordinate system, then every other co-ordinate system K' is a Galileian one, when, in relation to K, it is in a condition of uniform motion of translation. Relative to K' the mechanical laws of Galilei-Newton hold good exactly as they do with respect to K.

We advance a step farther in our generalisation when we express the tenet thus: If, relative to K, K' is a uniformly moving co-ordinate system devoid of rotation, then natural phenomena run their course with respect to K' according to exactly the same general laws as with respect to K. This statement is called the *principle of relativity* (in the restricted sense).

As long as one was convinced that all natural phenomena were capable of representation with the help of classical mechanics, there was no need to doubt the validity of this principle of relativity. But in view of the more recent development of electrodynamics and optics it became more and more evident that classical mechanics affords an insufficient foundation for the physical description of all natural phenomena. At this juncture the question of the validity of the principle of relativity became ripe for discussion, and it did not appear impossible that the answer to this question might be in the negative.

Nevertheless, there are two general facts which at the outset speak very much in favour of the validity of the principle of relativity. Even though classical mechanics does not supply us with a sufficiently broad basis for the theoretical presentation of all physical phenomena, still we must grant it a considerable measure of "truth," since it supplies us with the actual motions of the heavenly bodies with a delicacy of detail little short of wonderful. The principle of relativity must therefore apply with great accuracy in the domain of *mechanics*. But that a principle of such broad generality should hold with such exactness in one domain of phenomena, and yet should be invalid for another, is *a priori* not very probable.

We now proceed to the second argument, to which, moreover, we shall return later. If the principle of relativity (in the restricted sense) does not hold, then the Galileian co-ordinate systems K, K', K'', etc., which are moving uniformly relative to each other, will not be *equivalent* for the description of natural phenomena. In this case we should be constrained to believe that natural laws are capable of being formulated in a particularly simple manner, and of course only on condition that, from amongst all possible Galileian co-ordinate systems, we should have chosen *one* (K_0) of a particular state of motion as our body of reference. We should then be justified (because of its merits for the description of natural phenomena) in calling this system "absolutely at rest," and all other Galileian systems K "in motion." If, for instance, our embankment were the system K_0, then our railway carriage would be a system K, relative to which less simple laws would hold than with respect to K_0. This diminished simplicity would be due to the fact that the carriage K would be in motion (*i.e.* "really") with respect to K_0. In the general laws of nature which have been formulated with reference to K, the magnitude and direction of the velocity of the carriage would necessarily play a part. We should expect, for instance, that the note emitted by an organ-pipe placed with its axis parallel to the direction of travel would be different from that emitted if the axis of the pipe were placed perpendicular to this direction. Now in virtue of its motion in an orbit round the sun, our earth is comparable with a railway carriage travelling with a velocity of about 30 kilometres per second. If the principle of relativity were not valid we should therefore expect that the direction of motion of the earth at any moment would enter into the laws of nature, and also that physical systems in their behaviour would be dependent on the orientation in space with respect to the earth. For owing to the alteration in direction of the velocity of revolution of the earth in the course of a year, the earth cannot be at rest relative to the hypothetical system K_0 throughout the whole year. However, the most careful observations have never revealed such anisotropic properties in terrestrial physical space, *i.e.* a physical non-equivalence of different directions. This is very powerful argument in favour of the principle of relativity.

SIX

The Theorem of the Addition of Velocities Employed in Classical Mechanics

Let us suppose our old friend the railway carriage to be travelling along the rails with a constant velocity v, and that a man traverses the length of the carriage in the direction of travel with a velocity w. How quickly or, in other words, with what velocity W does the man advance relative to the embankment during the process? The only possible answer seems to result from the following consideration: If the man were to stand still for a second, he would advance relative to the embankment through a distance v equal numerically to the velocity of the carriage. As a consequence of his walking, however, he traverses an additional distance w relative to the carriage, and hence also relative to the embankment, in this second, the distance w being numerically equal to the velocity with which he is walking. Thus in total he covers the distance $W = v + w$ relative to the embankment in the second considered. We shall see later that this result, which expresses the theorem of the addition of velocities employed in classical mechanics, cannot be maintained; in other words, the law that we have just written down does not hold in reality. For the time being, however, we shall assume its correctness.

SEVEN

The Apparent Incompatibility of the Law of Propagation of Light with the Principle of Relativity

There is hardly a simpler law in physics than that according to which light is propagated in empty space. Every child at school knows, or believes he knows, that this propagation takes place in straight lines with a velocity c = 300,000 km./sec. At all events we know with great exactness that this velocity is the same for all colours, because if this were not the case, the minimum of emission would not be observed simultaneously for different colours during the eclipse of a fixed star by its dark neighbour. By means of similar considerations based on observations of double stars, the Dutch astronomer De Sitter was also able to show that the velocity of propagation of light cannot depend on the velocity of motion of the body emitting the light. The assumption that this velocity of propagation is dependent on the direction "in space" is in itself improbable.

In short, let us assume that the simple law of the constancy of the velocity of light c (in vacuum) is justifiably believed by the child at school. Who would imagine that this simple law has plunged the conscientiously thoughtful physicist into the greatest intellectual difficulties? Let us consider how these difficulties arise.

Of course we must refer the process of the propagation of light (and indeed every other process) to a rigid reference-body (co-ordinate system). As such a system let us again choose our embankment. We shall imagine the air above it to have been removed. If a ray of light be sent along the embankment, we see from the above that the tip of the ray will be transmitted with the velocity c relative to the embankment. Now let us suppose that our railway carriage is again travelling along the railway lines with the velocity v, and that its direction is the same as that of the ray of light, but its velocity of course much less. Let us inquire about the velocity of propagation of the ray of light relative to the carriage. It is obvious that we can here apply the consideration of the previous section, since the ray of light plays the part of the man walking along relatively to the carriage. The velocity W of the man relative to the embankment is here replaced by the velocity of light relative to the embankment. w is the required velocity of light with respect to the carriage, and we have

$$w = c - v.$$

The velocity of propagation of a ray of light relative to the carriage thus comes out smaller than c.

But this result comes into conflict with the principle of relativity set forth in Section 5. For, like every other general law of nature, the law of the transmission of light *in vacuo* must, according to the principle of relativity, be the same for the railway carriage as reference-body as when the rails are the body of reference. But, from our above consideration, this would appear to be impossible. If every ray of light is propagated relative to the embankment with the velocity *c*, then for this reason it would appear that another law of propagation of light must necessarily hold with respect to the carriage—a result contradictory to the principle of relativity.

In view of this dilemma there appears to be nothing else for it than to abandon either the principle of relativity or the simple law of the propagation of light *in vacuo*. Those of you who have carefully followed the preceding discussion are almost sure to expect that we should retain the principle of relativity, which appeals so convincingly to the intellect because it is so natural and simple. The law of the propagation of light *in vacuo* would then have to be replaced by a more complicated law comformable to the principle of relativity. The development of theoretical physics shows, however, that we cannot pursue this course. The epoch-making theoretical investigations of H. A. Lorentz on the electrodynamical and optical phenomena connected with moving bodies show that experience in this domain leads conclusively to a theory of electromagnetic phenomena, of which the law of the constancy of the velocity of light *in vacuo* is a necessary consequence. Prominent theoretical physicists were therefore more inclined to reject the principle of relativity, in spite of the fact that no empirical data had been found which were contradictory to this principle.

At this juncture the theory of relativity entered the arena. As a result of an analysis of the physical conceptions of time and space, it became evident that *in reality there is not the least incompatibility between the principle of relativity and the law of propagation of light*, and that by systematically holding fast to both these laws a logically rigid theory could be arrived at. This theory has been called the *special theory of relativity* to distinguish it from the extended theory, with which we shall deal later. In the following pages we shall present the fundamental ideas of the special theory of relativity.

EIGHT

On the Idea of Time in Physics

Lightning has struck the rails on our railway embankment at two places *A* and *B* far distant from each other. I make the additional assertion that these two lightning flashes occurred simultaneously. If I ask you whether there is sense in this statement, you will answer my question with a decided "Yes." But if I now approach you with the request to explain to me the sense of the statement more precisely, you find after some consideration that the answer to this question is not so easy as it appears at first sight.

After some time perhaps the following answer would occur to you: "The significance of the statement is clear in itself and needs no further explanation; of course it would require some consideration if I were to be commissioned to determine by observations whether in the actual case the two events took place simultaneously or not." I cannot be satisfied with this answer for the following reason. Supposing that as a result of ingenious consideration an able meteorologist were to discover that the lightning must always strike the places *A* and *B* simultaneously, then we should be faced with the task of testing whether or not this theoretical result is in accordance with the reality. We encounter the same difficulty with all physical statements in which the conception "simultaneous" plays a part. The concept does not exist for the physicist until he has the possibility of discovering whether or not it is fulfilled in an actual case. We thus require a definition of simultaneity such that this definition supplies us with the method by means of which, in the present case, he can decide by experiment whether or not both the lightning strokes occurred simultaneously. As long as this requirement is not satisfied, I allow myself to be deceived as a physicist (and of course the same applies if I am not a physicist), when I imagine that I am able to attach a meaning to the statement of simultaneity. (I would ask the reader not to proceed farther until he is fully convinced on this point.)

After thinking the matter over for some time you then offer the following suggestion with which to test simultaneity. By measuring along the rails, the connecting line *AB* should be measured up and an observer placed at the mid-point *M* of the distance *AB*. This observer should be supplied with an arrangement (*e.g.* two mirrors inclined at 90°) which allows him visually to observe both places *A* and *B* at the same time. If the observer perceives the two flashes of lightning at the same time, then they are simultaneous.

I am very pleased with this suggestion, but for all that I cannot regard the matter as quite settled, because I feel constrained to raise the following objection: "Your definition would certainly be right, if only I knew that the light by means of which the observer at *M* perceives the lightning flashes travels along the length $A \longrightarrow M$ with the same velocity as along the length $B \longrightarrow M$. But an examination of this supposition would only be possible if we already had at our disposal the means of measuring time. It would thus appear as though we were moving here in a logical circle."

After further consideration you cast a somewhat disdainful glance at me—and rightly so—and you declare: "I maintain my previous definition nevertheless, because in reality it assumes absolutely nothing about light. There is only *one* demand to be made of the definition of simultaneity, namely, that in every real case it must supply us with an empirical decision as to whether or not the conception that has to be defined is fulfilled. That my definition satisfies this demand is indisputable. That light requires the same time to traverse the path $A \longrightarrow M$ as for the path $B \longrightarrow M$ is in reality neither a *supposition nor a hypothesis* about the physical nature of light, but a *stipulation* which I can make of my own freewill in order to arrive at a definition of simultaneity."

It is clear that this definition can be used to give an exact meaning not only to *two* events, but to as many events as we care to choose, and independently of the positions of the scenes of the events with respect to the body of reference[1] (here the railway embankment). We are thus led also to a definition of "time" in physics. For this purpose we suppose

[1] We suppose further, that, when three events *A*, *B* and *C* occur in different places in such a manner that *A* is simultaneous with *B*, and *B* is simultaneous with *C* (simultaneous in the sense of the above definition), then the criterion for the simultaneity of the pair of events *A*, *C* is also satisfied. This assumption is a physical hypothesis about the law of propagation of light; it must certainly be fulfilled if we are to maintain the law of the constancy of the velocity of light *in vacuo*.

that clocks of identical construction are placed at the points *A*, *B* and *C* of the railway line (co-ordinate system), and that they are set in such a manner that the positions of their pointers are simultaneously (in the above sense) the same. Under these conditions we understand by the "time" of an event the reading (position of the hands) of that one of these clocks which is in the immediate vicinity (in space) of the event. In this manner a time-value is associated with every event which is essentially capable of observation.

This stipulation contains a further physical hypothesis, the validity of which will hardly be doubted without empirical evidence to the contrary. It has been assumed that all these clocks go *at the same rate* if they are of identical construction. Stated more exactly: When two clocks arranged at rest in different places of a reference-body are set in such a manner that a *particular* position of the pointers of the one clock is *simultaneous* (in the above sense) with the *same* position of the pointers of the other clock, then identical "settings" are always simultaneous (in the sense of the above definition).

NINE

The Relativity of Simultaneity

Up to now our considerations have been referred to a particular body of reference, which we have styled a "railway embankment." We suppose a very long train travelling along the rails with the constant velocity v and in the direction indicated in Fig. 1. People travelling in this train

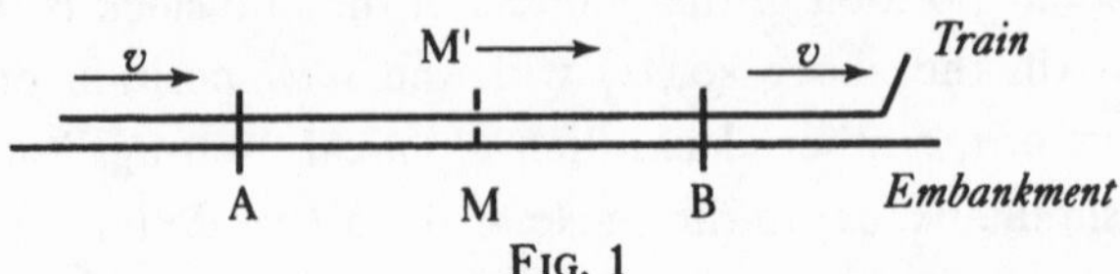

FIG. 1

will with advantage use the train as a rigid reference-body (co-ordinate system); they regard all events in reference to the train. Then every event which takes place along the line also takes place at a particular point of the train. Also the definition of simultaneity can be given relative to the train in exactly the same way as with respect to the embankment. As a natural consequence, however, the following question arises:

Are two events (*e.g.* the two strokes of lightning A and B) which are simultaneous *with reference to the railway embankment* also simultaneous ***relatively to the train?*** We shall show directly that the answer must be in the negative.

When we say that the lightning strokes *A* and *B* are simultaneous with respect to the embankment, we mean: the rays of light emitted at the places *A* and *B*, where the lightning occurs, meet each other at the mid-point *M* of the length *A* ⟶ *B* of the embankment. But the events *A* and *B* also correspond to positions *A* and *B* on the train. Let *M′* be the mid-point of the distance *A* ⟶ *B* on the travelling train. Just when the flashes[1] of lightning occur, this point *M′* naturally coincides with the point *M*, but it moves towards the right in the diagram with the velocity v of the train. If an observer sitting in the position *M′* in the train did not possess this velocity, then he would remain permanently at *M*, and the light rays emitted by the flashes of lightning *A* and *B* would reach him simultaneously, *i.e.* they would meet just where he is situated. Now in reality (considered with reference to the railway embankment) he is hastening towards the beam of light coming from *B*, whilst he is riding on ahead of the beam of light coming from *A*. Hence the observer will see the beam of light emitted from *B* earlier than he will see that emitted from *A*. Observers who take the railway train as their reference-body must therefore come to the conclusion that the lightning flash *B* took place earlier than the lightning flash *A*. We thus arrive at the important result:

Events which are simultaneous with reference to the embankment are not simultaneous with respect to the train, and

[1] As judged from the embankment.

vice versa (relativity of simultaneity). Every reference-body (co-ordinate system) has its own particular time; unless we are told the reference-body to which the statement of time refers, there is no meaning in a statement of the time of an event.

Now before the advent of the theory of relativity it had always tacitly been assumed in physics that the statement of time had an absolute significance, *i.e.* that it is independent of the state of motion of the body of reference. But we have just seen that this assumption is incompatible with the most natural definition of simultaneity; if we discard this assumption, then the conflict between the law of the propagation of light *in vacuo* and the principle of relativity (developed in Section 7) disappears.

We were led to that conflict by the considerations of Section 6, which are now no longer tenable. In that section we concluded that the man in the carriage, who traverses the distance *w per second* relative to the carriage, traverses the same distance also with respect to the embankment *in each second* of time. But, according to the foregoing considerations, the time required by a particular occurrence with respect to the carriage must not be considered equal to the duration of the same occurrence as judged from the embankment (as reference-body). Hence it cannot be contended that the man in walking travels the distance *w* relative to the railway line in a time which is equal to one second as judged from the embankment.

Moreover, the considerations of Section 6 are based on yet a second assumption, which, in the light of a strict consideration, appears to be arbitrary, although it was always tacitly made even before the introduction of the theory of relativity.

TEN

On the Relativity of the Conception of Distance

Let us consider two particular points on the train[1] travelling along the embankment with the velocity v, and inquire as to their distance apart. We already know that it is necessary to have a body of reference for the measurement of a distance, with respect to which body the distance can be measured up. It is the simplest plan to use the train itself as reference-body (co-ordinate system). An observer in the train measures the interval by marking off his measuring-rod in a straight line (*e.g.* along the floor of the carriage) as many times as is necessary to take him from the one marked point to the other. Then the number which tells us how often the rod has to be laid down is the required distance.

It is a different matter when the distance has to be judged from the railway line. Here the following method suggests itself. If we call A' and B' the two points on the train whose distance apart is required, then both of these points are mov-

[1] *E.g.* the middle of the first and of the twentieth carriage.

ing with the velocity v along the embankment. In the first place we require to determine the points A and B of the embankment which are just being passed by the two points A' and B' at a particular time t—judged from the embankment. These points A and B of the embankment can be determined by applying the definition of time given in Section 8. The distance between these points A and B is then measured by repeated application of the measuring-rod along the embankment.

A priori it is by no means certain that this last measurement will supply us with the same result as the first. Thus the length of the train as measured from the embankment may be different from that obtained by measuring in the train itself. This circumstance leads us to a second objection which must be raised against the apparently obvious consideration of Section 6. Namely, if the man in the carriage covers the distance w in a unit of time—*measured from the train*—then this distance—*as measured from the embankment*—is not necessarily also equal to w.

ELEVEN

The Lorentz Transformation

The results of the last three sections show that the apparent incompatibility of the law of propagation of light with the principle of relativity (Section 7) has been derived by means of a consideration which borrowed two unjustifiable hypotheses from classical mechanics; these are as follows:

(1) The time-interval (time) between two events is independent of the condition of motion of the body of reference.

(2) The space-interval (distance) between two points of a rigid body is independent of the condition of motion of the body of reference.

If we drop these hypotheses, then the dilemma of Section 7 disappears, because the theorem of the addition of velocities derived in Section 6 becomes invalid. The possibility presents itself that the law of the propagation of light *in vacuo* may be compatible with the principle of relativity, and the question arises: How have we to modify the considerations of Section 6 in order to remove the apparent disagreement between these two fundamental results of experience? This question leads to a general one. In the discussion of Section 6 we have to do with places and times relative both to the train and to the embankment. How are we to find the place and time of an event in relation to the train, when we know the place and time of the event with respect to the railway embankment? Is there a thinkable answer to this question of such a nature that the law of transmission of light *in vacuo* does not contradict the principle of relativity? In other words: Can we conceive of a relation between place and time of the individual events relative to both reference-bodies, such that every ray of light possesses the velocity of transmission c relative to the embankment and relative to the train? This question leads to a quite definite positive answer, and to a perfectly definite transformation law for the space-time magnitudes of an event when changing over from one body of reference to another.

Before we deal with this, we shall introduce the following incidental consideration. Up to the present we have only considered events taking place along the embankment, which had mathematically to assume the function of a straight line. In the manner indicated in Section 2 we can imagine this reference-body supplemented laterally and in a vertical direction by means of a framework of rods, so that an event which takes place anywhere can be localised with reference to this framework. Similarly, we can imagine the train travelling with the velocity v to be continued across the whole of space, so that every event, no matter how far off it may be, could also be localised with respect to the second framework. Without committing any fundamental error, we can disregard the fact that in reality these frameworks would continually interfere with each other, owing to the impenetrability of solid bodies. In every such framework we imagine three surfaces perpendicular to each other marked out, and designated as "co-ordinate planes" ("co-ordinate system"). A co-ordinate system K then corresponds to the embankment, and a co-ordinate system K' to the train. An event, wherever it may have taken place, would be fixed in space with respect to K by the three perpendiculars x, y, z on the co-ordinate planes, and with regard to time by a time-value t. Relative to K', *the same event* would be fixed in respect of space and time by corresponding values x', y', z', t',which of course are not identical with x, y, z, t. It has already been set forth in detail how these magnitudes are to be regarded as results of physical measurements.

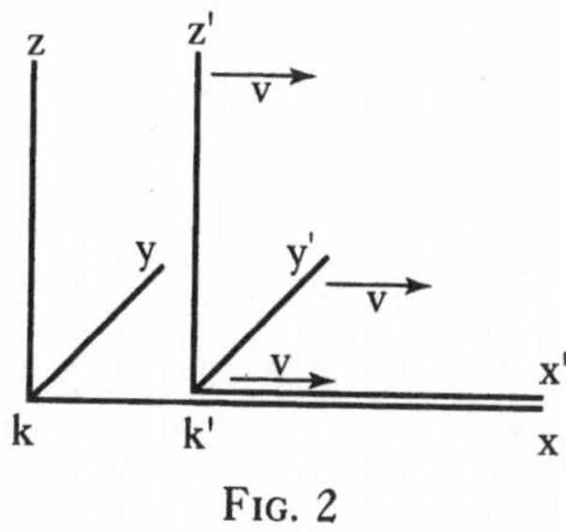

FIG. 2

Obviously our problem can be exactly formulated in the following manner. What are the values x', y', z', t', of an event with respect to K', when the magnitudes x, y, z, t, of the same event with respect to K are given? The relations must be so chosen that the law of the transmission of light *in vacuo* is satisfied for one and the same ray of light (and of course for every ray) with respect to K and K'. For the relative orientation in space of the co-ordinate systems indicated in the diagram (Fig. 2), this problem is solved by means of the equations:

$$x' = \frac{x - vt}{\sqrt{1 - \frac{v^2}{c^2}}}$$

$$y' = y$$

$$z' = z$$

$$t' = \frac{t - \frac{v}{c^2} \cdot x}{\sqrt{1 - \frac{v^2}{c^2}}}.$$

This system of equations is known as the "Lorentz transformation."[1]

If in place of the law of transmission of light we had taken as our basis the tacit assumptions of the older mechanics as to the absolute character of times and lengths, then instead of the above we should have obtained the following equations:

$$x' = x - vt$$

$$y' = y$$

$$z' = z$$

$$t' = t.$$

[1] A simple derivation of the Lorentz transformation is given in Appendix 1.

This system of equations is often termed the "Galilei transformation." The Galilei transformation can be obtained from the Lorentz transformation by substituting an infinitely large value for the velocity of light c in the latter transformation.

Aided by the following illustration, we can readily see that, in accordance with the Lorentz transformation, the law of the transmission of light *in vacuo* is satisfied both for the reference-body K and for the reference-body K'. A light-signal is sent along the positive x-axis, and this light-stimulus advances in accordance with the equation.

$$x = ct,$$

i.e. with the velocity c. According to the equations of the Lorentz transformation, this simple relation between x and t involves a relation between x' and t'. In point of fact, if we substitute for x the value ct in the first and fourth equations of the Lorentz transformation, we obtain:

$$x' = \frac{(c - v)t}{\sqrt{1 - \frac{v^2}{c^2}}}$$

$$t' = \frac{\left(1 - \frac{v}{c}\right)t}{\sqrt{1 - \frac{v^2}{c^2}}},$$

from which, by division, the expression

$$x' = ct'$$

immediately follows. If referred to the system K', the propagation of light takes place according to this equation. We thus see that the velocity of transmission relative to the reference-body K' is also equal to c. The same result is obtained for rays of light advancing in any other direction whatsoever. Of course this is not surprising, since the equations of the Lorentz transformation were derived conformably to this point of view.

TWELVE

The Behaviour of Measuring-Rods and Clocks in Motion

I place a metre-rod in the x'-axis of K' in such a manner that one end (the beginning) coincides with the point $x' = 0$, whilst the other end (the end of the rod) coincides with the point $x' = 1$. What is the length of the metre-rod relatively to the system K? In order to learn this, we need only ask where the beginning of the rod and the end of the rod lie with respect to K at a particular time t of the system K. By means of the first equation of the Lorentz transformation the values of these two points at the time $t = 0$ can be shown to be

$$x_{(\text{beginning of rod})} = 0 \sqrt{1 - \frac{v^2}{c^2}}$$

$$x_{(\text{end of rod})} = 1 . \sqrt{1 - \frac{v^2}{c^2}},$$

the distance between the points being $\sqrt{1 - \frac{v^2}{c^2}}$. But the metre-rod is moving with the velocity v relative to K. It therefore follows that the length of a rigid metre-rod moving in the direction of its length with a velocity v is $\sqrt{1 - v^2/c^2}$ of a metre. The rigid rod is thus shorter when in motion than when at rest, and the more quickly it is moving, the shorter is the rod. For the velocity $v = c$ we should have $\sqrt{1 - v^2/c^2 = 0}$, and for still greater velocities the square-root becomes imaginary. From this we conclude that in the theory of relativity the velocity c plays the part of a limiting velocity, which can neither be reached nor exceeded by any real body.

Of course this feature of the velocity c as a limiting velocity also clearly follows from the equations of the Lorentz transformation, for these become meaningless if we choose values of v greater than c.

If, on the contrary, we had considered a metre-rod at rest in the x-axis with respect to K, then we should have found that the length of the rod as judged from K' would have been $\sqrt{1 - v^2/c^2}$; this is quite in accordance with the principle of relativity which forms the basis of our considerations.

A priori it is quite clear that we must be able to learn something about the physical behaviour of measuring-rods and clocks from the equations of transformation, for the magnitudes x, y, z, t, are nothing more nor less than the results of measurements obtainable by means of measuring-rods and clocks. If we had based our considerations on the Galileian transformation we should not have obtained a contraction of the rod as a consequence of its motion.

Let us now consider a seconds-clock which is permanently

situated at the origin ($x' = 0$) of K'. $t' = 0$ and $t' = 1$ are two successive ticks of this clock. The first and fourth equations of the Lorentz transformation give for these two ticks:

$$t = 0$$

and

$$t = \frac{1}{\sqrt{1 - \frac{v^2}{c^2}}}$$

As judged from K, the clock is moving with the velocity v; as judged from this reference-body, the time which elapses between two strokes of the clock is not one second, but $\frac{1}{\sqrt{1 - \frac{v^2}{c^2}}}$ seconds, *i.e.* a somewhat larger time. As a consequence of its motion the clock goes more slowly than when at rest. Here also the velocity c plays the part of an unattainable limiting velocity.

THIRTEEN

Theorem of the Addition of the Velocities. The Experiment of Fizeau

Now in practice we can move clocks and measuring-rods only with velocities that are small compared with the velocity of light; hence we shall hardly be able to compare the results of the previous section directly with the reality. But, on the other hand, these results must strike you as being very singular, and for that reason I shall now draw another conclusion from the theory, one which can easily be derived from the foregoing considerations, and which has been most elegantly confirmed by experiment.

In Section 6 we derived the theorem of the addition of velocities in one direction in the form which also results from the hypotheses of classical mechanics. This theorem can also be deduced readily from the Galilei transformation (Section 11). In place of the man walking inside the carriage, we introduce a point moving relatively to the co-ordinate system K' in accordance with the equation

$$x = wt'.$$

By means of the first and fourth equations of the Galilei transformation we can express x' and t' in terms of x and t, and we then obtain

$$x = (v + w)t.$$

This equation expresses nothing else than the law of motion of the point with reference to the system K (of the man with reference to the embankment). We denote this velocity by the symbol W, and we then obtain, as in Section 6,

$$W = v + w \qquad . \quad . \quad . \qquad \text{(A)}.$$

But we can carry out this consideration just as well on the basis of the theory of relativity. In the equation

$$x' = wt'$$

we must then express x' and t' in terms of x and t, making use of the first and fourth equations of the *Lorentz transformation.* Instead of the equation (A) we then obtain the equation

$$W = \frac{v + w}{1 + \dfrac{vw}{c^2}} \qquad . \quad . \quad . \qquad \text{(B)},$$

which corresponds to the theorem of addition for velocities in one direction according to the theory of relativity. The question now arises as to which of these two theorems is the better in accord with experience. On this point we are enlightened by a most important experiment which the brilliant physicist Fizeau performed more than half a century ago, and which has been repeated since then by some of the best experimental physicists, so that there can be no doubt about its result. The experiment is concerned with the following question. Light travels in a motionless liquid with a particular velocity w. How quickly does it travel in the direction of the arrow in the tube T (see the accompanying diagram, Fig. 3) when the liquid above mentioned is flowing through the tube with a velocity v?

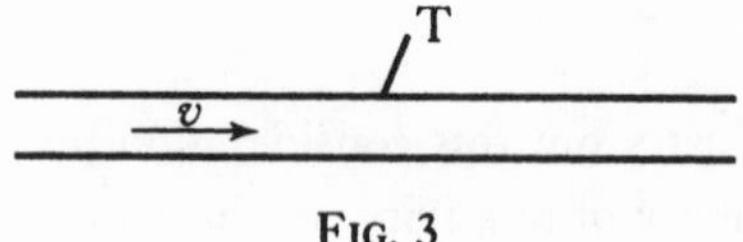

FIG. 3

In accordance with the principle of relativity we shall certainly have to take for granted that the propagation of light always takes place with the same velocity w *with respect to the liquid*, whether the latter is in motion with reference to other bodies or not. The velocity of light relative to the liquid and the velocity of the latter relative to the tube are thus known, and we require the velocity of light relative to the tube.

It is clear that we have the problem of Section 6 again before us. The tube plays the part of the railway embankment or of the co-ordinate system K, the liquid plays the part of the carriage or of the co-ordinate system K', and finally, the light plays the part of the man walking along the carriage, or of the moving point in the present section. If we denote the velocity of the light relative to the tube by W, then this is given by the equation (A) or (B), according as the Galilei transformation or the Lorentz transformation corresponds to the facts. Experiment[1] decides in favour of equation (B) derived from the theory of relativity, and the agreement is, indeed, very exact. According to recent and most excellent measurements by Zeeman, the influence of the velocity of flow v on the propagation of light is represented by formula (B) to within one per cent.

Nevertheless we must now draw attention to the fact that a theory of this phenomenon was given by H. A. Lorentz long before the statement of the theory of relativity. This theory was of a purely electrodynamical nature, and was obtained by the use of particular hypotheses as to the electromagnetic structure of matter. This circumstance, however, does not in the least diminish the conclusiveness of the experiment as a crucial test in favour of the theory of relativity, for the electrodynamics of Maxwell-Lorentz, on which the original theory was based, in no way opposes the theory of relativity. Rather has the latter been developed from electrodynamics as an astoundingly simple combination of generalisation of the hypotheses, formerly independent of each other, on which electrodynamics was built.

[1] Fizeau found $W = w + v\left(1 - \frac{1}{n^2}\right)$, where $n = \frac{c}{w}$ is the index of refraction of the liquid. On the other hand, owing to the smallness of $\frac{vw}{c^2}$ as compared with 1, we can replace (B) in the first place by $W = (w + v)\left(1 - \frac{vw}{c^2}\right)$, or to the same order of approximation by $w + v\left(1 - \frac{1}{n^2}\right)$, which agrees with Fizeau's result.

FOURTEEN

The Heuristic Value of the Theory of Relativity

Our train of thought in the foregoing pages can be epitomised in the following manner. Experience has led to the conviction that, on the one hand, the principle of relativity holds true and that on the other hand the velocity of transmission of light *in vacuo* has to be considered equal to a constant c. By uniting these two postulates we obtained the law of transformation for the rectangular co-ordinates x, y, z and the time t of the events which constitute the processes of nature. In this connection we did not obtain the Galilei transformation, but, differing from classical mechanics, the *Lorentz transformation*.

The law of transmission of light, the acceptance of which is justified by our actual knowledge, played an important part in this process of thought. Once in possession of the Lorentz transformation, however, we can combine this with the principle of relativity, and sum up the theory thus:

Every general law of nature must be so constituted that it is transformed into a law of exactly the same form when, instead of the space-time variables x, y, z, t of the original co-ordinate system K, we introduce new space-time variables x', y', z', t' of a co-ordinate system K'. In this connection the relation between the ordinary and the accented magnitudes is given by the Lorentz transformation. Or in brief: General laws of nature are co-variant with respect to Lorentz transformations.

This is a definite mathematical condition that the theory of relativity demands of a natural law, and in virtue of this, the theory becomes a valuable heuristic aid in the search for general laws of nature. If a general law of nature were to be found which did not satisfy this condition, then at least one of the two fundamental assumptions of the theory would have been disproved. Let us now examine what general results the latter theory has hitherto evinced.

FIFTEEN

General Results of the Theory

It is clear from our previous considerations that the (special) theory of relativity has grown out of electrodynamics and optics. In these fields it has not appreciably altered the predictions of theory, but it has considerably simplified the theoretical structure, *i.e.* the derivation of laws, and—what is incomparably more important—it has considerably reduced the number of independent hypotheses forming the basis of theory. The special theory of relativity has rendered the Maxwell-Lorentz theory so plausible, that the latter would have been generally accepted by physicists even if experiment had decided less unequivocally in its favour.

Classical mechanics required to be modified before it could come into line with the demands of the special theory of relativity. For the main part, however, this modification affects only the laws for rapid motions, in which the velocities of matter v are not very small as compared with the velocity of light. We have experience of such rapid motions only in the case of electrons and ions; for other motions the variations from the laws of classical mechanics are too small to make themselves evident in practice. We shall not consider the motion of stars until we come to speak of the general theory of relativity. In accordance with the theory of relativity the kinetic energy of a material point of mass m is no longer given by the well-known expression

$$m\frac{v^2}{2},$$

but by the expression

$$\frac{mc^2}{\sqrt{1-\frac{v^2}{c^2}}}.$$

This expression approaches infinity as the velocity v approaches the velocity of light c. The velocity must therefore always remain less than c, however great may be the energies used to produce the acceleration. If we develop the expression for the kinetic energy in the form of a series, we obtain

$$mc^2 + m\frac{v^2}{2} + \frac{3}{8}m\frac{v^4}{c^2} + \ldots.$$

When $\frac{v^2}{c^2}$ is small compared with unity, the third of these terms is always small in comparison with the second, which last is alone considered in classical mechanics. The first term

mc^2 does not contain the velocity, and requires no consideration if we are only dealing with the question as to how the energy of a point-mass depends on the velocity. We shall speak of its essential significance later.

The most important result of a general character to which the special theory of relativity has led is concerned with the conception of mass. Before the advent of relativity, physics recognised two conservation laws of fundamental importance, namely, the law of the conservation of energy and the law of the conservation of mass; these two fundamental laws appeared to be quite independent of each other. By means of the theory of relativity they have been united into one law. We shall now briefly consider how this unification came about, and what meaning is to be attached to it.

The principle of relativity requires that the law of the conservation of energy should hold not only with reference to a co-ordinate system K, but also with respect to every co-ordinate system K' which is in a state of uniform motion of translation relative to K, or, briefly, relative to every "Galileian" system of co-ordinates. In contrast to classical mechanics, the Lorentz transformation is the deciding factor in the transition from one such system to another.

By means of comparatively simple considerations we are led to draw the following conclusion from these premises, in conjunction with the fundamental equations of the electrodynamics of Maxwell: A body moving with the velocity v, which absorbs[1] an amount of energy E_0 in the form of radiation

[1] E_0 is the energy taken up, as judged from a co-ordinate system moving with the body.

without suffering an alteration in velocity in the process, has, as a consequence, its energy increased by an amount

$$\frac{E_0}{\sqrt{1-\frac{v^2}{c^2}}}.$$

In consideration of the expression given above for the kinetic energy of the body, the required energy of the body comes out to be

$$\frac{\left(m+\frac{E_0}{c^2}\right)c^2}{\sqrt{1-\frac{v^2}{c^2}}}.$$

Thus the body has the same energy as a body of mass $\left(m+\frac{E_0}{c^2}\right)$ moving with the velocity v. Hence we can say: If a body takes up an amount of energy E_0, then its inertial mass increases by an amount $\frac{E_0}{c^2}$; the inertial mass of a body is not a constant, but varies according to the change in the energy of the body. The inertial mass of a system of bodies can even be regarded as a measure of its energy. The law of the conservation of the mass of a system becomes identical with the law of the conservation of energy, and is only valid provided that the system neither takes up nor sends out energy. Writing the expression for the energy in the form

$$\frac{mc^2 + E_0}{\sqrt{1 - \frac{v^2}{c^2}}},$$

we see that the term mc^2, which has hitherto attracted our attention, is nothing else than the energy possessed by the body[1] before it absorbed the energy E_0.

A direct comparison of this relation with experiment is not possible at the present time (1920; see Note, p. 53), owing to the fact that the changes in energy E_0 to which we can subject a system are not large enough to make themselves perceptible as a change in the inertial mass of the system. $\frac{E_0}{c^2}$ is too small in comparison with the mass m, which was present before the alteration of the energy. It is owing to this circumstance that classical mechanics was able to establish successfully the conservation of mass as a law of independent validity.

Let me add a final remark of a fundamental nature. The success of the Faraday-Maxwell interpretation of electromagnetic action at a distance resulted in physicists becoming convinced that there are no such things as instantaneous actions at a distance (not involving an intermediary medium) of the type of Newton's law of gravitation. According to the theory of relativity, action at a distance with the velocity of light always takes the place of instantaneous action at a distance or of action at a distance with an infinite velocity of transmission. This is connected with the fact that the velocity c plays a fundamental rôle in this theory. In Part II we shall see in what

[1] As judged from a co-ordinate system moving with the body.

way this result becomes modified in the general theory of relativity.

NOTE.—With the advent of nuclear transformation processes, which result from the bombardment of elements by α-particles, protons, deuterons, neutrons or γ-rays, the equivalence of mass and energy expressed by the relation $E = mc^2$ has been amply confirmed. The sum of the reacting masses, together with the mass equivalent of the kinetic energy of the bombarding particle (or photon), is always greater than the sum of the resulting masses. The difference is the equivalent mass of the kinetic energy of the particles generated, or of the released electromagnetic energy (γ-photons). In the same way, the mass of a spontaneously disintegrating radioactive atom is always greater than the sum of the masses of the resulting atoms by the mass equivalent of the kinetic energy of the particles generated (or of the photonic energy). Measurements of the energy of the rays emitted in nuclear reactions, in combination with the equations of such reactions, render it possible to evaluate atomic weights to a high degree of accuracy.

R. W. L.

SIXTEEN

Experience and the Special Theory of Relativity

To what extent is the special theory of relativity supported by experience? This question is not easily answered for the reason already mentioned in connection with the fundamental experiment of Fizeau. The special theory of relativity has crystallised out from the Maxwell-Lorentz theory of electromagnetic phenomena. Thus all facts of experience which support the electromagnetic theory also support the theory of relativity. As being of particular importance, I mention here the fact that the theory of relativity enables us to predict the effects produced on the light reaching us from the fixed stars. These results are obtained in an exceedingly simple manner, and the effects indicated, which are due to the relative motion of the earth with reference to those fixed stars, are found to be in accord with experience. We refer to the yearly movement of the apparent position of the fixed stars resulting from the motion of the earth round the sun (aberration), and to the influence of the radial components of the relative motions of the fixed stars with respect to the earth on the colour of the light reaching us from them. The latter effect manifests itself in a slight displacement of the spectral lines of the light transmitted to us from a fixed star, as compared with the position of the same spectral lines when they are produced by a terrestrial source of light (Doppler principle). The experimental arguments in favour of the Maxwell-Lorentz theory, which are at the same time arguments in favour of the theory of relativity, are too numerous to be set forth here. In reality they limit the theoretical possibilities to such an extent, that no other theory than that of Maxwell and Lorentz has been able to hold its own when tested by experience.

But there are two classes of experimental facts hitherto obtained which can be represented in the Maxwell-Lorentz theory only by the introduction of an auxiliary hypothesis, which in itself—*i.e.* without making use of the theory of relativity—appears extraneous.

It is known that cathode rays and the so-called β-rays emitted by radioactive substances consist of negatively electrified particles (electrons) of very small inertia and large velocity. By examining the deflection of these rays under the influence of electric and magnetic fields, we can study the law of motion of these particles very exactly.

In the theoretical treatment of these electrons, we are faced with the difficulty that electrodynamic theory of itself is unable to give an account of their nature. For since electrical masses of one sign repel each other, the negative electrical masses constituting the electron would necessarily be scattered under the influence of their mutual repulsions, unless there are forces of another kind operating between them, the nature of which has hitherto remained obscure to us.[1] If we now assume that the relative distances between the electrical masses constituting the electron remain unchanged during the motion of the electron (rigid connection in the sense of classical mechanics), we arrive at a law of motion of the electron which does not agree with experience. Guided by purely formal points of view, H. A. Lorentz was the first to introduce the hypothesis that the form of the electron experiences a contraction in the direction of motion in consequence of that motion, the contracted length being proportional to the expession $\sqrt{1-\frac{v^2}{c^2}}$. This hypothesis, which is not justifiable by any electrodynamical facts, supplies us then with that particular law of motion which has been confirmed with great precision in recent years.

The theory of relativity leads to the same law of motion, without requiring any special hypothesis whatsoever as to the structure and the behaviour of the electron. We arrived at a similar conclusion of Section 13 in connection with the experiment of Fizeau, the result of which is foretold by the theory of relativity without the necessity of drawing on hypotheses as to the physical nature of the liquid.

The second class of facts to which we have alluded has reference to the question whether or not the motion of the earth in space can be made perceptible in terrestrial experiments. We have already remarked in Section 5 that all attempts of this

[1] The general theory of relativity renders it likely that the electrical masses of an electron are held together by gravitational forces.

nature led to a negative result. Before the theory of relativity was put forward, it was difficult to become reconciled to this negative result, for reasons now to be discussed. The inherited prejudices about time and space did not allow any doubt to arise as to the prime importance of the Galileian transformation for changing over from one body of reference to another. Now assuming that the Maxwell-Lorentz equations hold for a reference-body *K*, we then find that they do not hold for a reference-body *K′* moving uniformly with respect to *K*, if we assume that the relations of the Galileian transformation exist between the co-ordinates of *K* and *K′*. It thus appears that, of all Galileian co-ordinate systems, one (*K*) corresponding to a particular state of motion is physically unique. This result was interpreted physically by regarding *K* as at rest with respect to a hypothetical æther of space. On the other hand, all co-ordinate systems *K′* moving relatively to *K* were to be regarded as in motion with respect to the æther. To this motion of *K′* against the æther ("æther-drift" relative to *K′*) were attributed the more complicated laws which were supposed to hold relative to *K′*. Strictly speaking, such an æther-drift ought also to be assumed relative to the earth, and for a long time the efforts of physicists were devoted to attempts to detect the existence of an æther-drift at the earth's surface.

In one of the most notable of these attempts Michelson devised a method which appears as though it must be decisive. Imagine two mirrors so arranged on a rigid body that the reflecting surfaces face each other. A ray of light requires a perfectly definite time *T* to pass from one mirror to the other and back again, if the whole system be at rest with respect to the æther. It is found by calculation, however, that a slightly different time T' is required for this process, if the body, together with the mirrors, be moving relatively to the æther. And yet another point: it is shown by calculation that for a given velocity v with reference to the æther, this time T' is different when the body is moving perpendicularly to the planes of the mirrors from that resulting when the motion is parallel to these planes. Although the estimated difference between these two times is exceedingly small, Michelson and Morley performed an experiment involving interference in which this difference should have been clearly detectable. But the experiment gave a negative result—a fact very perplexing to physicists. Lorentz and FitzGerald rescued the theory from this difficulty by assuming that the motion of the body relative to the æther produces a contraction of the body in the direction of motion, the amount of contraction being just sufficient to compensate for the difference in time mentioned above. Comparison with the discussion in Section 12 shows that also from the standpoint of the theory of relativity this solution of the difficulty was the right one. But on the basis of the theory of relativity the method of interpretation is incomparably more satisfactory. According to this theory there is no such thing as a "specially favoured" (unique) co-ordinate system to occasion the introduction of the æther-idea, and hence there can be no æther-drift, nor any experiment with which to demonstrate it. Here the contraction of moving bodies follows from the two fundamental principles of the theory, without the introduction of particular hypotheses; and as the prime factor involved in this contraction we find, not the motion in itself, to which we cannot attach any meaning, but the motion with respect to the body of reference chosen in the particular case in point. Thus for a co-ordinate system moving with the earth the mirror system of Michelson and Morley is not shortened, but it *is* shortened for a co-ordinate system which is at rest relatively to the sun.

SEVENTEEN

Minkowski's Four-Dimensional Space

The non-mathematician is seized by a mysterious shuddering when he hears of "four-dimensional" things, by a feeling not unlike that awakened by thoughts of the occult. And yet there is no more common-place statement than that the world in which we live is a four-dimensional space-time continuum.

Space is a three-dimensional continuum. By this we mean that it is possible to describe the position of a point (at rest) by means of three numbers (co-ordinates) x, y, z, and that there is an indefinite number of points in the neighbourhood of this one, the position of which can be described by co-ordinates such as x_1, y_1, z_1, which may be as near as we choose to the respective values of the co-ordinates x, y, z of the first point. In virtue of the latter property we speak of a "continuum," and owing to the fact that there are three co-ordinates we speak of it as being "three-dimensional."

Similarly, the world of physical phenomena which was briefly called "world" by Minkowski is naturally four-dimensional in the space-time sense. For it is composed of individual events, each of which is described by four numbers, namely, three space co-ordinates x, y, z and a time co-ordinate, the time-value t. The "world" is in this sense also a continuum; for to every event there are as many "neighbouring" events (realised or at least thinkable) as we care to choose, the co-ordinates x_1, y_1, z_1, t_1 of which differ by an indefinitely small amount from those of the event x, y, z, t originally considered. That we have not been accustomed to regard the world in this sense as a four-dimensional continuum is due to the fact that in physics, before the advent of the theory of relativity, time played a different and more independent rôle, as compared with the space co-ordinates. It is for this reason that we have been in the habit of treating time as an independent continuum. As a matter of fact, according to classical mechanics, time is absolute, *i.e.* it is independent of the position and the condition of motion of the system of co-ordinates. We see this expressed in the last equation of the Galileian transformation ($t' = t$).

The four-dimensional mode of consideration of the "world" is natural on the theory of relativity, since according to this theory time is robbed of its independence. This is shown by the fourth equation of the Lorentz transformation:

$$t' = \frac{t - \frac{v}{c^2}x}{\sqrt{1 - \frac{v^2}{c^2}}}.$$

Moreover, according to this equation the time difference $\Delta t'$ of two events with respect to K' does not in general vanish, even when the time difference Δt of the same events with reference to K vanishes. Pure "space-distance" of two events with respect to K results in "time-distance" of the same events with respect to K'. But the discovery of Minkowski, which was of importance for the formal development of the theory of relativity, does not lie here. It is to be found rather in the fact of his recognition that the four-dimensional space-time continuum of the theory of relativity, in its most essential formal properties, shows a pronounced relationship to the three-dimensional continuum of Euclidean geometrical space.[1] In order to give due prominence to this relationship, however, we must replace the usual time co-ordinate t by an imaginary magnitude $\sqrt{-1} \cdot ct$ proportional to it. Under these conditions, the natural laws satisfying the demands of the (special) theory of relativity assume mathematical forms, in which the time co-ordinate plays exactly the same rôle as the three space co-ordinates. Formally, these four co-ordinates correspond exactly to the three space co-ordinates in Euclidean geometry. It must be clear even to the non-mathematician that, as a consequence of this purely formal addition to our knowledge, the theory perforce gained clearness in no mean measure.

These inadequate remarks can give the reader only a vague notion of the important idea contributed by Minkowski. Without it the general theory of relativity, of which the fundamental ideas are developed in the following pages, would perhaps have got no farther than its long clothes. Minkowski's work is doubtless difficult of access to anyone inexperienced in math-

[1] Cf. the somewhat more detailed discussion in Appendix 2.

ematics, but since it is not necessary to have a very exact grasp of this work in order to understand the fundamental ideas of either the special or the general theory of relativity, I shall leave it here at present, and revert to it only towards the end of Part II.

Part II

The General Theory of Relativity

EIGHTEEN

Special and General Principle of Relativity

The basal principle, which was the pivot of all our previous considerations, was the *special* principle of relativity, *i.e.* the principle of the physical relativity of all *uniform* motion. Let us once more analyse its meaning carefully.

It was at all times clear that, from the point of view of the idea it conveys to us, every motion must be considered only as a relative motion. Returning to the illustration we have frequently used of the embankment and the railway carriage, we can express the fact of the motion here taking place in the following two forms, both of whch are equally justifiable:

(*a*) The carriage is in motion relative to the embankment.

(*b*) The embankment is in motion relative to the carriage.

In (*a*) the embankment, in (*b*) the carriage, serves as the body of reference in our statement of the motion taking place. If it is simply a question of detecting or of describing the motion involved, it is in principle immaterial to what reference-body we refer the motion. As already mentioned, this is self-evident, but it must not be confused with the much more comprehensive statement called "the principle of relativity," which we have taken as the basis of our investigations.

The principle we have made use of not only maintains that we may equally well choose the carriage or the embankment as our reference-body for the description of any event (for this, too, is self-evident). Our principle rather asserts what follows: If we formulate the general laws of nature as they are obtained from experience, by making use of

(*a*) the embankment as reference-body,

(*b*) the railway carriage as reference-body,

then these general laws of nature (*e.g.* the laws of mechanics or the law of the propagation of light *in vacuo*) have exactly the same form in both cases. This can also be expressed as follows: For the *physical* description of natural processes, neither of the reference-bodies K, K' is unique (lit. "specially marked out") as compared with the other. Unlike the first, this latter statement need not of necessity hold *a priori*; it is not contained in the conceptions of "motion" and "reference-body" and derivable from them; only *experience* can decide as to its correctness or incorrectness.

Up to the present, however, we have by no means maintained the equivalence of *all* bodies of reference K in connection with the formulation of natural laws. Our course was more on the following lines. In the first place, we started out from the assumption that there exists a reference-body K, whose condition of motion is such that the Galileian law holds with respect to it: A particle left to itself and sufficiently far removed from all other particles moves uniformly in a straight line. With reference to K (Galileian reference-body) the laws of nature were to be as simple as possible. But in addition to *K*, all bodies of reference *K′* should be given preference in this sense, and they should be exactly equivalent to *K* for the formulation of natural laws, provided that they are in a state of *uniform rectilinear and non-rotary motion* with respect to *K*; all these bodies of reference are to be regarded as Galileian reference-bodies. The validity of the principle of relativity was assumed only for these reference-bodies, but not for others (*e.g.* those possessing motion of a different kind). In this sense we speak of the *special* principle of relativity, or special theory of relativity.

In contrast to this we wish to understand by the "general principle of relativity" the following statement: All bodies of reference *K*, *K′*, etc., are equivalent for the description of natural phenomena (formulation of the general laws of nature), whatever may be their state of motion. But before proceeding farther, it ought to be pointed out that this formulation must be replaced later by a more abstract one, for reasons which will become evident at a later stage.

Since the introduction of the special principle of relativity has been justified, every intellect which strives after generalisation must feel the temptation to venture the step towards the general principle of relativity. But a simple and apparently quite reliable consideration seems to suggest that, for the present at any rate, there is little hope of success in such an attempt. Let us imagine ourselves transferred to our old friend the railway carriage, which is travelling at a uniform rate. As long as it is moving uniformly, the occupant of the carriage is not sensible of its motion, and it is for this reason that he can without reluctance interpret the facts of the case as indicating that the carriage is at rest, but the embankment in motion. Moreover, according to the special principle of relativity, this interpretation is quite justified also from a physical point of view.

If the motion of the carriage is now changed into a non-uniform motion, as for instance by a powerful application of the brakes, then the occupant of the carriage experiences a correspondingly powerful jerk forwards. The retarded motion is manifested in the mechanical behaviour of bodies relative to the person in the railway carriage. The mechanical behaviour is different from that of the case previously considered, and for this reason it would appear to be impossible that the same mechanical laws hold relatively to the non-uniformly moving carriage, as hold with reference to the carriage when at rest or in uniform motion. At all events it is clear that the Galileian law does not hold with respect to the non-uniformly moving carriage. Because of this, we feel compelled at the present juncture to grant a kind of absolute physical reality to non-uniform motion, in opposition to the general principle of relativity. But in what follows we shall soon see that this conclusion cannot be maintained.

NINETEEN

The Gravitational Field

"If we pick up a stone and then let it go, why does it fall to the ground?" The usual answer to this question is: "Because it is attracted by the earth." Modern physics formulates the answer rather differently for the following reason. As a result of the more careful study of electromagnetic phenomena, we have come to regard action at a distance as a process impossible without the intervention of some intermediary medium. If, for instance, a magnet attracts a piece of iron, we cannot be content to regard this as meaning that the magnet acts directly on the iron through the intermediate empty space, but we are constrained to imagine—after the manner of Faraday—that the magnet always calls into being something physically real in the space around it, that something being what we call a "magnetic field." In its turn this magnetic field operates on the piece of iron, so that the latter strives to move towards the magnet. We shall not discuss here the justification for this incidental conception, which is indeed a somewhat arbitrary one. We shall only mention that with its aid electromagnetic phenomena can be theoretically represented much more satisfactorily than without it, and this applies particularly to the transmission of electromagnetic waves. The effects of gravitation also are regarded in an analogous manner.

The action of the earth on the stone takes place indirectly. The earth produces in its surroundings a gravitational field, which acts on the stone and produces its motion of fall. As we know from experience, the intensity of the action on a body diminishes according to a quite definite law, as we proceed farther and farther away from the earth. From our point of view this means: The law governing the properties of the gravitational field in space must be a perfectly definite one, in order correctly to represent the diminution of gravitational action with the distance from operative bodies. It is something like this: The body (*e.g.* the earth) produces a field in its immediate neighbourhood directly; the intensity and direction of the field at points farther removed from the body are thence determined by the law which governs the properties in space of the gravitational fields themselves.

In contrast to electric and magnetic fields, the gravitational field exhibits a most remarkable property, which is of fundamental importance for what follows. Bodies which are moving under the sole influence of a gravitational field receive an acceleration, ***which does not in the least depend either on the material or on the physical state of the body***. For instance, a piece of lead and a piece of wood fall in exactly the same manner in a gravitational field (*in vacuo*), when they start off from rest or with the same initial velocity. This law, which holds most accurately, can be expressed in a different form in the light of the following consideration.

According to Newton's law of motion, we have

$$\text{(Force)} = \text{(inertial mass)} \times \text{(acceleration)},$$

where the "inertial mass" is a characteristic constant of the accelerated body. If now gravitation is the cause of the acceleration, we then have

$$\text{(Force)} = \text{(gravitational mass)} \times \text{(intensity of the gravitational field)},$$

where the "gravitational mass" is likewise a characteristic constant for the body. From these two relations follows:

$$\text{(acceleration)} = \frac{\text{gravitational mass)}}{\text{(inertial mass)}} \text{(intensity of the gravitational field)}.$$

If now, as we find from experience, the acceleration is to be independent of the nature and the condition of the body and always the same for a given gravitational field, then the ratio of the gravitational to the inertial mass must likewise be the same for all bodies. By a suitable choice of units we can thus make this ratio equal to unity. We then have the following law: The *gravitational* mass of a body is equal to its *inertial* mass.

It is true that this important law had hitherto been recorded in mechanics, but it had not been *interpreted.* A satisfactory interpretation can be obtained only if we recognise the following fact: *The same* quality of a body manifests itself according to circumstances as "inertia" or as "weight" (lit. "heaviness"). In the following section we shall show to what extent this is actually the case, and how this question is connected with the general postulate of relativity.

TWENTY

The Equality of Inertial and Gravitational Mass as an Argument for the General Postulate of Relativity

We imagine a large portion of empty space, so far removed from stars and other appreciable masses, that we have before us approximately the conditions required by the fundamental law of Galilei. It is then possible to choose a Galileian reference-body for this part of space (world), relative to which points at rest remain at rest and points in motion continue permanently in uniform rectilinear motion. As reference-body let us imagine a spacious chest resembling a room with an observer inside who is equipped with apparatus. Gravitation naturally does not exist for this observer. He must fasten himself with strings to the floor, otherwise the slightest impact against the floor will cause him to rise slowly towards the ceiling of the room.

To the middle of the lid of the chest is fixed externally a hook with rope attached, and now a "being" (what kind of a being is immaterial to us) begins pulling at this with a constant force. The chest together with the observer then begins to move "upwards" with a uniformly accelerated motion. In course of time their velocity will reach unheard-of values—

provided that we are viewing all this from another reference-body which is not being pulled with a rope.

But how does the man in the chest regard the process? The acceleration of the chest will be transmitted to him by the reaction of the floor of the chest. He must therefore take up this pressure by means of his legs if he does not wish to be laid out full length on the floor. He is then standing in the chest in exactly the same way as anyone stands in a room of a house on our earth. If he releases a body which he previously had in his hand, the acceleration of the chest will no longer be transmitted to this body, and for this reason the body will approach the floor of the chest with an accelerated relative motion. The observer will further convince himself *that the acceleration of the body towards the floor of the chest is always of the same magnitude, whatever kind of body he may happen to use for the experiment.*

Relying on his knowledge of the gravitational field (as it was discussed in the preceding section), the man in the chest will thus come to the conclusion that he and the chest are in a gravitational field which is constant with regard to time. Of course he will be puzzled for a moment as to why the chest does not fall in this gravitational field. Just then, however, he discovers the hook in the middle of the lid of the chest and the rope which is attached to it, and he consequently comes to the conclusion that the chest is suspended at rest in the gravitational field.

Ought we to smile at the man and say that he errs in his conclusion? I do not believe we ought to if we wish to remain consistent; we must rather admit that his mode of grasping the situation violates neither reason nor known mechanical laws.

Even though it is being accelerated with respect to the "Galileian space" first considered, we can nevertheless regard the chest as being at rest. We have thus good grounds for extending the principle of relativity to include bodies of reference which are accelerated with respect to each other, and as a result we have gained a powerful argument for a generalised postulate of relativity.

We must note carefully that the possibility of this mode of interpretation rests on the fundamental property of the gravitational field of giving all bodies the same acceleration, or, what comes to the same thing, on the law of the equality of inertial and gravitational mass. If this natural law did not exist, the man in the accelerated chest would not be able to interpret the behaviour of the bodies around him on the supposition of a gravitational field, and he would not be justified on the grounds of experience in supposing his reference-body to be "at rest."

Suppose that the man in the chest fixes a rope to the inner side of the lid, and that he attaches a body to the free end of the rope. The result of this will be to stretch the rope so that it will hang "vertically" downwards. If we ask for an opinion of the cause of tension in the rope, the man in the chest will say: "The suspended body experiences a downward force in the gravitational field, and this is neutralised by the tension of the rope; what determines the magnitude of the tension of the rope is the *gravitational mass* of the suspended body." On the other hand, an observer who is poised freely in space will interpret the condition of things thus: "The rope must perforce take part in the accelerated motion of the chest, and it transmits this motion to the body attached to it. The tension of the rope is just large enough to effect the acceleration of the body. That which determines the magnitude of the tension of the rope is the *inertial mass* of the body." Guided by this example, we see that our extension of the principle of relativity implies the *necessity* of the law of the equality of inertial and gravitational mass. Thus we have obtained a physical interpretation of this law.

From our consideration of the accelerated chest we see that a general theory of relativity must yield important results on the laws of gravitation. In point of fact, the systematic pursuit of the general idea of relativity has supplied the laws satisfied by the gravitational field. Before proceeding farther, however, I must warn the reader against a misconception suggested by these considerations. A gravitational field exists for the man in the chest, despite the fact that there was no such field for the co-ordinate system first chosen. Now we might easily suppose that the existence of a gravitational field is always only an *apparent* one. We might also think that, regardless of the kind of gravitational field which may be present, we could always choose another reference-body such that *no* gravitational field exists with reference to it. This is by no means true for all gravitational fields, but only for those of quite special form. It is, for instance, impossible to choose a body of reference such that, as judged from it, the gravitational field of the earth (in its entirety) vanishes.

We can now appreciate why that argument is not convincing, which we brought forward against the general principle of relativity at the end of Section 18. It is certainly true that the observer in the railway carriage experiences a jerk forwards as a result of the application of the brake, and that he recognises in this the non-uniformity of motion (retardation) of the carriage. But he is compelled by nobody to refer this jerk to a "real" acceleration (retardation) of the carriage. He might also interpret his experience thus: "My body of reference (the carriage) remains permanently at rest. With reference to it, however, there exists (during the period of application of the brakes) a gravitational field which is directed forwards and which is variable with respect to time. Under the influence of this field, the embankment together with the earth moves nonuniformly in such a manner that their original velocity in the backwards direction is continuously reduced."

TWENTY-ONE

In What Respects Are the Foundations of Classical Mechanics and of the Special Theory of Relativity Unsatisfactory?

We have already stated several times that classical mechanics starts out from the following law: Material particles sufficiently far removed from other material particles continue to move uniformly in a straight line or continue in a state of rest. We have also repeatedly emphasised that this fundamental law can only be valid for bodies of reference *K* which possess certain unique states of motion, and which are in uniform translational motion relative to each other. Relative to other reference-bodies *K* the law is not valid. Both in classical mechanics and in the special theory of relativity we therefore differentiate between reference-bodies *K* relative to which the recognised "laws of nature" can be said to hold, and reference-bodies *K* relative to which these laws do not hold.

But no person whose mode of thought is logical can rest satisfied with this condition of things. He asks: "How does it come that certain reference-bodies (or their states of motion) are given priority over other reference-bodies (or their states of motion)? *What is the reason for this preference?* In order to show clearly what I mean by this question, I shall make use of a comparison.

I am standing in front of a gas range. Standing alongside of each other on the range are two pans so much alike that one may be mistaken for the other. Both are half full of water. I notice that steam is being emitted continuously from the one pan, but not from the other. I am surprised at this, even if I have never seen either a gas range or a pan before. But if I now notice a luminous something of bluish colour under the first pan but not under the other, I cease to be astonished, even if I have never before seen a gas flame. For I can only say that this bluish something will cause the emission of the steam, or at least *possibly* it may do so. If, however, I notice the bluish something in neither case, and if I observe that the one continuously emits steam whilst the other does not, then I shall remain astonished and dissatisfied until I have discovered some circumstance to which I can attribute the different behaviour of the two pans.

Analogously, I seek in vain for a real something in classical mechanics (or in the special theory of relativity) to which I can attribute the different behaviour of bodies considered with respect to the reference-systems K and K'.[1] Newton saw this objection and attempted to invalidate it, but without success. But E. Mach recognised it most clearly of all, and because of

[1] The objection is of importance more especially when the state of motion of the reference-body is of such a nature that it does not require any external agency for its maintenance, *e.g.* in the case when the reference-body is rotating uniformly.

this objection he claimed that mechanics must be placed on a new basis. It can only be got rid of by means of a physics which is comformable to the general principle of relativity, since the equations of such a theory hold for every body of reference, whatever may be its state of motion.

TWENTY-TWO

A Few Inferences from the General Principle of Relativity

The considerations of Section 20 show that the general principle of relativity puts us in a position to derive properties of the gravitational field in a purely theoretical manner. Let us suppose, for instance, that we know the space-time "course" for any natural process whatsoever, as regards the manner in which it takes place in the Galileian domain relative to a Galileian body of reference *K*. By means of purely theoretical operations (*i.e.* simply by calculation) we are then able to find how this known natural process appears, as seen from a reference-body *K′* which is accelerated relatively to *K*. But since a gravitational field exists with respect to this new body of reference *K′*, our consideration also teaches us how the gravitational field influences the process studied.

For example, we learn that a body which is in a state of uniform rectilinear motion with respect to *K* (in accordance with the law of Galilei) is executing an accelerated and in general curvilinear motion with respect to the accelerated reference-body *K′* (chest). This acceleration or curvature corresponds to the influence on the moving body of the gravitational field prevailing relatively to *K'*. It is known that a gravitational field influences the movement of bodies in this way, so that our consideration supplies us with nothing essentially new.

However, we obtain a new result of fundamental importance when we carry out the analogous consideration for a ray of light. With respect to the Galileian reference-body *K*, such a ray of light is transmitted rectilinearly with the velocity *c*. It can easily be shown that the path of the same ray of light is no longer a straight line when we consider it with reference to the accelerated chest (reference-body *K'*). From this we conclude, ***that, in general, rays of light are propagated curvilinearly in gravitational fields.*** In two respects this result is of great importance.

In the first place, it can be compared with the reality. Although a detailed examination of the question shows that the curvature of light rays required by the general theory of relativity is only exceedingly small for the gravitational fields at our disposal in practice, its estimated magnitude for light rays passing the sun at grazing incidence is nevertheless 1.7 seconds of arc. This ought to manifest itself in the following way. As seen from the earth, certain fixed stars appear to be in the neighbourhood of the sun, and are thus capable of observation during a total eclipse of the sun. At such times, these stars ought to appear to be displaced outwards from the sun by an amount indicated above, as compared with their apparent position in the sky when the sun is situated at another part of the heavens. The examination of the correctness or otherwise of this deduction is a problem of the greatest importance, the early solution of which is to be expected of astronomers.[1]

In the second place our result shows that, according to the general theory of relativity, the law of the constancy of the velocity of light *in vacuo*, which constitutes one of the two fundamental assumptions in the special theory of relativity and to which we have already frequently referred, cannot claim any unlimited validity. A curvature of rays of light can only take place when the velocity of propagation of light varies with position. Now we might think that as a consequence of this, the special theory of relativity and with it the whole theory of relativity would be laid in the dust. But in reality this is not the case. We can only conclude that the special theory of relativity cannot claim an unlimited domain of validity; its results hold only so long as we are able to disregard the influences of gravitational fields on the phenomena (*e.g.* of light).

Since it has often been contended by opponents of the theory of relativity that the special theory of relativity is overthrown by the general theory of relativity, it is perhaps advisable to make the facts of the case clearer by means of an appropriate comparison. Before the development of electrodynamics the laws of electrostatics were looked upon as the laws of electricity. At the present time we know that electric fields can be derived correctly from electrostatic considerations only for the case, which is never strictly realised, in which the electrical masses are quite at rest relatively to each

[1] By means of the star photographs of two expeditions equipped by a Joint Committee of the Royal and Royal Astronomical Societies, the existence of the deflection of light demanded by theory was first confirmed during the solar eclipse of 29th May, 1919. (Cf. Appendix 3.)

other, and to the co-ordinate system. Should we be justified in saying that for this reason electrostatics is overthrown by the field-equations of Maxwell in electrodynamics? Not in the least. Electrostatics is contained in electrodynamics as a limiting case; the laws of the latter lead directly to those of the former for the case in which the fields are invariable with regard to time. No fairer destiny could be allotted to any physical theory, than that it should of itself point out the way to the introduction of a more comprehensive theory, in which it lives on as a limiting case.

In the example of the transmission of light just dealt with, we have seen that the general theory of relativity enables us to derive theoretically the influence of a gravitational field on the course of natural processes, the laws of which are already known when a gravitational field is absent. But the most attractive problem, to the solution of which the general theory of relativity supplies the key, concerns the investigation of the laws satisfied by the gravitational field itself. Let us consider this for a moment.

We are acquainted with space-time domains which behave (approximately) in a "Galileian" fashion under suitable choice of reference-body, *i.e.* domains in which gravitational fields are absent. If we now refer such a domain to a reference-body K' possessing any kind of motion, then relative to K' there exists a gravitational field which is variable with respect to space and time.[1] The character of this field will of course depend on the motion chosen for K'. According to the general theory of rel-

[1] This follows from a generalisation of the discussion in Section 20.

ativity, the general law of the gravitational field must be satisfied for all gravitational fields obtainable in this way. Even though by no means all gravitational fields can be produced in this way, yet we may entertain the hope that the general law of gravitation will be derivable from such gravitational fields of a special kind. This hope has been realised in the most beautiful manner. But between the clear vision of this goal and its actual realisation it was necessary to surmount a serious difficulty, and as this lies deep at the root of things, I dare not withhold it from the reader. We require to extend our ideas of the space-time continuum still farther.

TWENTY-THREE

Behaviour of Clocks and Measuring-Rods on a Rotating Body of Reference

Hitherto I have purposely refrained from speaking about the physical interpretation of space- and time-data in the case of the general theory of relativity. As a consequence, I am guilty of a certain slovenliness of treatment, which, as we know from the special theory of relativity, is far from being unimportant and pardonable. It is now high time that we remedy this defect; but I would mention at the outset, that this matter lays no small claims on the patience and on the power of abstraction of the reader.

We start off again from quite special cases, which we have frequently used before. Let us consider a space-time domain in which no gravitational field exists relative to a reference-body K whose state of motion has been suitably chosen. K is then a Galileian reference-body as regards the domain considered, and the results of the special theory of relativity hold relative to K. Let us suppose the same domain referred to a second body of reference K', which is rotating uniformly with respect to K. In order to fix our ideas, we shall imagine K' to be in the form of a plane circular disc, which rotates uniformly in its own plane about its centre. An observer who is sitting eccentrically on the disc K' is sensible of a force which acts outwards in a radial direction, and which would be interpreted as an effect of inertia (centrifugal force) by an observer who was at rest with respect to the original reference-body K. But the observer on the disc may regard his disc as a reference-body which is "at rest"; on the basis of the general principle of relativity he is justified in doing this. The force acting on himself, and in fact on all other bodies which are at rest relative to the disc, he regards as the effect of a gravitational field. Nevertheless, the space-distribution of this gravitational field is of a kind that would not be possible on Newton's theory of gravitation.[1] But since the observer believes in the general theory of relativity, this does not disturb him; he is quite in the right when he believes that a general law of gravitation can be formulated—a law which not only explains the motion of the stars correctly, but also the field of force experienced by himself.

The observer performs experiments on his circular disc with clocks and measuring-rods. In doing so, it is his intention to arrive at exact definitions for the signification of time- and space-data with reference to the circular disc K', these definitions being based on his observations. What will be his experience in this enterprise?

To start with, he places one of two identically constructed clocks at the centre of the circular disc, and the other on the

[1] The field disappears at the centre of the disc and increases proportionally to the distance from the centre as we proceed outwards.

edge of the disc, so that they are at rest relative to it. We now ask ourselves whether both clocks go at the same rate from the standpoint of the non-rotating Galileian reference-body *K*. As judged from this body, the clock at the centre of the disc has no velocity, whereas the clock at the edge of the disc is in motion relative to *K* in consequence of the rotation. According to a result obtained in Section 12, it follows that the latter clock goes at a rate permanently slower than that of the clock at the centre of the circular disc, *i.e.* as observed from *K*. It is obvious that the same effect would be noted by an observer whom we will imagine sitting alongside his clock at the centre of the circular disc. Thus on our circular disc, or, to make the case more general, in every gravitational field, a clock will go more quickly or less quickly, according to the position in which the clock is situated (at rest). For this reason it is not possible to obtain a reasonable definition of time with the aid of clocks which are arranged at rest with respect to the body of reference. A similar difficulty presents itself when we attempt to apply our earlier definition of simultaneity in such a case, but I do not wish to go any farther into this question.

Moreover, at this stage the definition of the space coordinates also presents insurmountable difficulties. If the observer applies his standard measuring-rod (a rod which is short as compared with the radius of the disc) tangentially to the edge of the disc, then, as judged from the Galileian system, the length of this rod will be less than 1, since, according to Section 12, moving bodies suffer a shortening in the direction of the motion. On the other hand, the measuring-rod will not experience a shortening in length, as judged from *K*, if it is applied to the disc in the direction of the radius. If, then, the observer first measures the circumference of the disc with his measuring-rod and then the diameter of the disc, on dividing the one by the other, he will not obtain as quotient the familiar number $\pi = 3 \cdot 14 \ldots$, but a larger number,[1] whereas of course, for a disc which is at rest with respect to *K*, this operation would yield π exactly. This proves that the propositions of Euclidean geometry cannot hold exactly on the rotating disc, nor in general in a gravitational field, at least if we attribute the length 1 to the rod in all positions and in every orientation. Hence the idea of a straight line also loses its meaning. We are therefore not in a position to define exactly the co-ordinates x, y, z relative to the disc by means of the method used in discussing the special theory, and as long as the co-ordinates and times of events have not been defined, we cannot assign an exact meaning to the natural laws in which these occur.

Thus all our previous conclusions based on general relativity would appear to be called in question. In reality we must make a subtle detour in order to be able to apply the postulate of general relativity exactly. I shall prepare the reader for this in the following paragraphs.

[1] Throughout this consideration we have to use the Galileian (non-rotating) system *K* as reference-body, since we may only assume the validity of the results of the special theory of relativity relative to *K* (relative to *K'* a gravitational field prevails).

TWENTY-FOUR

Euclidean and Non-Euclidean Continuum

The surface of a marble table is spread out in front of me. I can get from any one point on this table to any other point by passing continuously from one point to a "neighbouring" one, and repeating this process a (large) number of times, or, in other words, by going from point to point without executing "jumps." I am sure the reader will appreciate with sufficient clearness what I mean here by "neighbouring" and by "jumps" (if he is not too pedantic). We express this property of the surface by describing the latter as a continuum.

Let us now imagine that a large number of little rods of equal length have been made, their lengths being small compared with the dimensions of the marble slab. When I say they are of equal length, I mean that one can be laid on any other without the ends overlapping. We next lay four of these little rods on the marble slab so that they constitute a quadrilateral figure (a square), the diagonals of which are equally long. To ensure the equality of the diagonals, we make use of a little testing-rod. To this square we add similar ones, each of which has one rod in common with the first. We proceed in like manner with each of these squares until finally the whole marble slab is laid out with squares. The arrangement is such, that each side of a square belongs to two squares and each corner to four squares.

It is a veritable wonder that we can carry out this business without getting into the greatest difficulties. We only need to think of the following. If at any moment three squares meet at a corner, then two sides of the fourth square are already laid, and, as a consequence, the arrangement of the remaining two sides of the square is already completely determined. But I am now no longer able to adjust the quadrilateral so that its diagonals may be equal. If they are equal of their own accord, then this is an especial favour of the marble slab and of the little rods, about which I can only be thankfully surprised. We must experience many such surprises if the construction is to be successful.

If everything has really gone smoothly, then I say that the points of the marble slab constitute a Euclidean continuum with respect to the little rod, which has been used as a "distance" (line-interval). By choosing one corner of a square as "origin," I can characterize every other corner of a square with reference to this origin by means of two numbers. I only need state how many rods I must pass over when, starting from the origin, I proceed towards the "right" and then "upwards," in order to arrive at the corner of the square under consideration. These two numbers are then the "Cartesian co-ordinates" of this corner with reference to the "Cartesian co-ordinate system" which is determined by the arrangement of little rods.

By making use of the following modification of this abstract experiment, we recognise that there must also be cases in which the experiment would be unsuccessful. We shall suppose that the rods "expand" by an amount proportional to the increase of temperature. We heat the central part of the marble slab, but not the periphery, in which case two of our little rods can still be brought into coincidence at every position on the table. But our construction of squares must necessarily come into disorder during the heating, because the little rods on the central region of the table expand, whereas those on the outer part do not.

With reference to our little rods—defined as unit lengths—the marble slab is no longer a Euclidean continuum, and we are also no longer in the position of defining Cartesian coordinates directly with their aid, since the above construction can no longer be carried out. But since there are other things which are not influenced in a similar manner to the little rods (or perhaps not at all) by the temperature of the table, it is possible quite naturally to maintain the point of view that the marble slab is a "Euclidean continuum." This can be done in a satisfactory manner by making a more subtle stipulation about the measurement or the comparison of lengths.

But if rods of every kind (*i.e.* of every material) were to behave *in the same way* as regards the influence of temperature when they are on the variably heated marble slab, and if we had no other means of detecting the effect of temperature than the geometrical behaviour of our rods in experiments analogous to the one described above, then our best plan would be to assign the distance *one* to two points on the slab, provided that the ends of one of our rods could be made to coincide with these two points; for how else should we define the distance without our proceeding being in the highest measure grossly arbitrary? The method of Cartesian co-ordinates must then be discarded, and replaced by another which does not assume the validity of Euclidean geometry for rigid bodies.[1] The reader will notice that the situation depicted here corresponds to the one brought about by the general postulate of relativity (Section 23).

[1] Mathematicians have been confronted with our problem in the following form. If we are given a surface (*e.g.* an ellipsoid) in Euclidean three-dimensional space, then there exists for this surface a two-dimensional geometry, just as much as for a plane surface. Gauss undertook the task of treating this two-dimensional geometry from first principles, without making use of the fact that the surface belongs to a Euclidean continuum of three dimensions. If we imagine constructions to be made with rigid rods *in the surface* (similar to that above with the marble slab), we should find that different laws hold for these from those resulting on the basis of Euclidean plane geometry. The surface is not a Euclidean continuum with respect to the rods, and we cannot define Cartesian co-ordinates *in the surface*. Gauss indicated the principles according to which we can treat the geometrical relationships in the surface, and thus pointed out the way to the method of Riemann of treating multi-dimensional, non-Euclidean *continua*. Thus it is that mathematicians long ago solved the formal problems to which we are led by the general postulate of relativity.

TWENTY-FIVE

Gaussian Co-ordinates

According to Gauss, this combined analytical and geometrical mode of handling the problem can be arrived at in the following way. We imagine a system of arbitrary curves (see Fig. 4) drawn on the surface of the table.

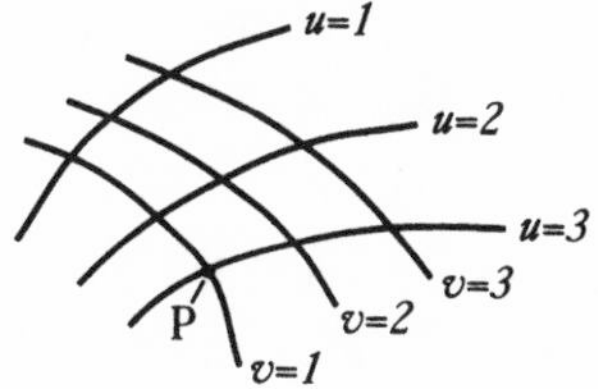

Fig. 4

These we designate as *u*-curves, and we indicate each of them by means of a number. The curves $u = 1$, $u = 2$ and $u = 3$ are drawn in the diagram. Between the curves $u = 1$ and $u = 2$ we must imagine an infinitely large number to be drawn, all of which correspond to real numbers lying between 1 and 2. We have then a system of *u*-curves, and this "infinitely dense" system covers the whole surface of the table. These *u*-curves must not intersect each other, and through each point of the surface one and only one curve must pass. Thus a perfectly definite value of u belongs to every point on the surface of the marble slab. In like manner we imagine a system of v-curves drawn on the surface. These satisfy the same conditions as the u-curves, they are provided with numbers in a corresponding manner, and they may likewise be of arbitrary shape. It follows that a value of u and a value of v belong to every point on the surface of the table. We call these two numbers the co-ordinates of the surface of the table (Gaussian co-ordinates). For example, the point P in the diagram has the Gaussian co-ordinates $u = 3, v = 1$. Two neighbouring points P and P' on the surface then correspond to the co-ordinates

$$P: \qquad u, v$$
$$P': \qquad u + du, v + dv,$$

where du and dv signify very small numbers. In a similar manner we may indicate the distance (line-interval) between P and P', as measured with a little rod, by means of the very small number ds. Then according to Gauss we have

$$ds^2 = g_{11}du^2 + 2g_{12}dudv + g_{22}\, dv^2,$$

where g_{11}, g_{12}, g_{22}, are magnitudes which depend in a perfectly definite way to u and v. The magnitudes g_{11}, g_{12} and g_{22} determine the behaviour of the rods relative to the u-curves and v-curves, and thus also relative to the surface of the table. For the case in which the points of the surface considered form a Euclidean continuum with reference to the measuring-rods, but only in this case, it is possible to draw the u-curves and v-curves and to attach numbers to them, in such a manner, that we simply have:

$$ds^2 = du^2 + dv^2.$$

Under these conditions, the u-curves and v-curves are straight lines in the sense of Euclidean geometry, and they are perpendicular to each other. Here the Gaussian co-ordinates are simply Cartesian ones. It is clear that Gauss co-ordinates are nothing more than an association of two sets of numbers with the points of the surface considered, of such a nature that numerical values differing very slightly from each other are associated with neighbouring points "in space."

So far, these considerations hold for a continuum of two dimensions. But the Gaussian method can be applied also to a continuum of three, four or more dimensions. If, for instance, a continuum of four dimensions be supposed available, we may represent it in the following way. With every point of the continuum we associate arbitrarily four numbers, x_1, x_2, x_3, x_4, which are known as "co-ordinates." Adjacent points correspond to adjacent values of the co-ordinates. If a distance ds is associated with the adjacent points P and P', this distance being measurable and well-defined from a physical point of view, then the following formula holds:

$$ds^2 = g_{11}dx_1{}^2 + 2g_{12}dx_1dx_2 \; . \; . \; . \; . + g_{44}dx_4{}^2,$$

where the magnitudes g_{11}, etc., have values which vary with the position in the continuum. Only when the continuum is a Euclidean one is it possible to associate the co-ordinates x_1 . . x_4 with the points of the continuum so that we have simply

$$ds^2 = dx_1^{\,2} + dx_2^{\,2} + dx_3^{\,2} + dx_4^{\,2}.$$

In this case relations hold in the four-dimensional continuum which are analogous to those holding in our three-dimensional measurements.

However, the Gauss treatment for ds^2 which we have given above is not always possible. It is only possible when sufficiently small regions of the continuum under consideration may be regarded as Euclidean continua. For example, this obviously holds in the case of the marble slab of the table and local variation of temperature. The temperature is practically constant for a small part of the slab, and thus the geometrical behaviour of the rods is *almost* as it ought to be according to the rules of Euclidean geometry. Hence the imperfections of the construction of squares in the previous section do not show themselves clearly until this construction is extended over a considerable portion of the surface of the table.

We can sum this up as follows: Gauss invented a method for the mathematical treatment of continua in general, in which "size-relations" ("distances" between neighbouring points) are defined. To every point of a continuum are assigned as many numbers (Gaussian co-ordinates) as the continuum has dimensions. This is done in such a way, that only one meaning can be attached to the assignment, and that numbers (Gaussian co-ordinates) which differ by an indefinitely small amount are assigned to adjacent points. The Gaussian co-ordinate system is a logical generalisation of the Cartesian co-ordinate system. It is also applicable to non-Euclidean continua, but only when, with respect to the defined "size" or "distance," small parts of the continuum under consideration behave more nearly like a Euclidean system, the smaller the part of the continuum under our notice.

TWENTY-SIX

The Space-Time Continuum of the Special Theory of Relativity Considered as a Euclidean Continuum

We are now in a position to formulate more exactly the idea of Minkowski, which was only vaguely indicated in Section 17. In accordance with the special theory of relativity, certain co-ordinate systems are given preference for the description of the four-dimensional, space-time continuum. We called these "Galileian co-ordinate systems." For these systems, the four co-ordinates x, y, z, t, which determine an event or—in other words—a point of the four-dimensional continuum, are defined physically in a simple manner, as set forth in detail in the first part of this book. For the transition from one Galileian system to another, which is moving uniformly with reference to the first, the equations of the Lorentz transformation are valid. These last form the basis for the derivation of deductions from the special theory of relativity, and in themselves they are nothing more than the expression of the universal validity of the law of transmission of light for all Galileian systems of reference.

Minkowski found that the Lorentz transformations satisfy the following simple conditions. Let us consider two neighbouring events, the relative position of which in the four-dimensional continuum is given with respect to a Galileian reference-body K by the space co-ordinate differences dx, dy, dz and the time-difference dt. With reference to a second Galileian system we shall suppose that the corresponding differences for these two events are dx', dy', dz', dt'. Then these magnitudes always fulfil the condition[1]

$$dx^2 + dy^2 + dz^2 - c^2dt^2 = dx'^2 + dy'^2 + dz'^2 - c^2dt'^2.$$

The validity of the Lorentz transformation follows from this condition. We can express this as follows: The magnitude

$$ds^2 = dx^2 + dy^2 + dz^2 - c^2dt^2,$$

which belongs to two adjacent points of the four-dimensional space-time continuum, has the same value for all selected (Galileian) reference-bodies. If we replace x, y, z $\sqrt{-1}\,ct$, by x_1, x_2, x_3, x_4, we also obtain the result that

$$ds^2 = dx_1{}^2 + dx_2{}^2 + dx_3{}^2 + dx_4{}^2$$

is independent of the choice of the body of reference. We call the magnitude ds the "distance" apart of the two events or four-dimensional points.

[1] Cf. Appendices 1 and 2. The relations which are derived there for the co-ordinates themselves are valid also for co-ordinate *differences*, and thus also for co-ordinate differentials (indefinitely small differences).

Thus, if we choose as time-variable the imaginary variable $\sqrt{-1}\,ct$ instead of the real quantity t, we can regard the space-time continuum—in accordance with the special theory of relativity—as a "Euclidean" four-dimensional continuum, a result which follows from the considerations of the preceding section.

TWENTY-SEVEN

The Space-Time Continuum of the General Theory of Relativity Is Not a Euclidean Continuum

In the first part of this book we were able to make use of space-time co-ordinates which allowed of a simple and direct physical interpretation, and which, according to Section 26, can be regarded as four-dimensional Cartesian co-ordinates. This was possible on the basis of the law of the constancy of the velocity of light. But according to Section 21, the general theory of relativity cannot retain this law. On the contrary, we arrived at the result that according to this latter theory the velocity of light must always depend on the co-ordinates when a gravitational field is present. In connection with a specific illustration in Section 23, we found that the presence of a gravitational field invalidates the definition of the co-ordinates and the time, which led us to our objective in the special theory of relativity.

In view of the results of these considerations we are led to the conviction that, according to the general principle of relativity, the space-time continuum cannot be regarded as a Euclidean one, but that here we have the general case, corresponding to the marble slab with local variations of temperature, and with which we made acquaintance as an example of a two-dimensional continuum. Just as it was there impossible to construct a Cartesian co-ordinate system from equal rods, so here it is impossible to build up a system (reference-body) from rigid bodies and clocks, which shall be of such a nature that measuring-rods and clocks, arranged rigidly with respect to one another, shall indicate position and time directly. Such was the essence of the difficulty with which we were confronted in Section 23.

But the considerations of Sections 25 and 26 show us the way to surmount this difficulty. We refer the four-dimensional space-time continuum in an arbitrary manner to Gauss co-ordinates. We assign to every point of the continuum (event) four numbers, x_1, x_2, x_3, x_4 (co-ordinates), which have not the least direct physical significance, but only serve the purpose of numbering the points of the continuum in a definite but arbitrary manner. This arrangement does not even need to be of such a kind that we must regard x_1, x_2, x_3 as "space" co-ordinates and x_4 as a "time" co-ordinate.

The reader may think that such a description of the world would be quite inadequate. What does it mean to assign to an event the particular co-ordinates x_1, x_2, x_3, x_4, if in themselves these co-ordinates have no significance? More careful consideration shows, however, that this anxiety is unfounded. Let us consider, for instance, a material point with any kind of motion. If this point had only a momentary existence without duration, then it would be described in space-time by a single system of values x_1, x_2, x_3, x_4. Thus its permanent existence must be characterised by an infinitely large number of such systems of values, the co-ordinate values of which are so close together as to give continuity; corresponding to the material point, we thus have a (uni-dimensional) line in the four-dimensional continuum. In the same way, any such lines in our continuum correspond to many points in motion. The only statements having regard to these points which can claim a physical existence are in reality the statements about their encounters. In our mathematical treatment, such an encounter is expressed in the fact that the two lines which represent the motions of the points in question have a particular system of co-ordinate values, x_1, x_2, x_3, x_4, in common. After mature consideration the reader will doubtless admit that in reality such encounters constitute the only actual evidence of a time-space nature with which we meet in physical statements.

When we were describing the motion of a material point relative to a body of reference, we stated nothing more than the encounters of this point with particular points of the reference-body. We can also determine the corresponding values of the time by the observation of encounters of the body with clocks, in conjunction with the observation of the encounter of the hands of clocks with particular points on the dials. It is just the same in the case of space-measurements by means of measuring-rods, as a little consideration will show.

The following statements hold generally: Every physical description resolves itself into a number of statements, each of which refers to the space-time coincidence of two events A and B. In terms of Gaussian co-ordinates, every such statement is expressed by the agreement of their four co-ordinates

x_1, x_2, x_3, x_4. Thus in reality, the description of the time-space continuum by means of Gauss co-ordinates completely replaces the description with the aid of a body of reference, without suffering from the defects of the latter mode of description; it is not tied down to the Euclidean character of the continuum which has to be represented.

TWENTY-EIGHT

Exact Formulation of the General Principle of Relativity

We are now in a position to replace the provisional formulation of the general principle of relativity given in Section 18 by an exact formulation. The form there used, "All bodies of reference *K*, *K′*, etc., are equivalent for the description of natural phenomena (formulation of the general laws of nature), whatever may be their state of motion," cannot be maintained, because the use of rigid reference-bodies, in the sense of the method followed in the special theory of relativity, is in general not possible in space-time description. The Gauss co-ordinate system has to take the place of the body of reference. The following statement corresponds to the fundamental idea of the general principle of relativity: "*All Gaussian co-ordinate systems are essentially equivalent for the formulation of the general laws of nature.*"

We can state this general principle of relativity in still another form, which renders it yet more clearly intelligible than it is when in the form of the natural extension of the special principle of relativity. According to the special theory of relativity, the equations which express the general laws of nature pass over into equations of the same form when, by making use of the Lorentz transformation, we replace the space-time variables x, y, z, t, of a (Galileian) reference-body K by the space-time variables x', y', z', t', of a new reference-body K'. According to the general theory of relativity, on the other hand, by application of ***arbitrary substitutions*** of the Gauss variables x_1, x_2, x_3, x_4, the equations must pass over into equations of the same form; for every transformation (not only the Lorentz transformation) corresponds to the transition of one Gauss co-ordinate system into another.

If we desire to adhere to our "old-time" three-dimensional view of things, then we can characterise the development which is being undergone by the fundamental idea of the general theory of relativity as follows: The special theory of relativity has reference to Galileian domains, *i.e.* to those in which no gravitational field exists. In this connection a Galileian reference-body serves as body of reference, *i.e.* a rigid body the state of motion of which is so chosen that the Galileian law of the uniform rectilinear motion of "isolated" material points holds relatively to it.

Certain considerations suggest that we should refer the same Galileian domains to ***non-Galileian*** reference-bodies also. A gravitational field of a special kind is then present with respect to these bodies (cf. Sections 20 and 23).

In gravitational fields there are no such things as rigid bodies with Euclidean properties; thus the fictitious rigid body of reference is of no avail in the general theory of relativity. The motion of clocks is also influenced by gravitational fields, and in such a way that a physical definition of time which is made directly with the aid of clocks has by no means the same degree of plausibility as in the special theory of relativity.

For this reason non-rigid reference-bodies are used, which are as a whole not only moving in any way whatsoever, but which also suffer alterations in form *ad lib.* during their motion. Clocks, for which the law of motion is of any kind, however irregular, serve for the definition of time. We have to imagine each of these clocks fixed at a point on the non-rigid reference-body. These clocks satisfy only the one condition, that the "readings" which are observed simultaneously on adjacent clocks (in space) differ from each other by an indefinitely small amount. This non-rigid reference-body, which might appropriately be termed a "reference-mollusc," is in the main equivalent to a Gaussian four-dimensional co-ordinate system chosen arbitrarily. That which gives the "mollusc" a certain comprehensibility as compared with the Gauss co-ordinate system is the (really unjustified) formal retention of the separate existence of the space co-ordinates as opposed to the time co-ordinate. Every point on the mollusc is treated as a space-point, and every material point which is at rest relatively to it as at rest, so long as the mollusc is considered as reference-body. The general principle of relativity requires that all these molluscs can be used as reference-bodies with equal right and equal success in the formulation of the general laws of nature; the laws themselves must be quite independent of the choice of mollusc.

The great power possessed by the general principle of relativity lies in the comprehensive limitation which is imposed on the laws of nature in consequence of what we have seen above.

TWENTY-NINE

The Solution of the Problem of Gravitation on the Basis of the General Principle of Relativity

If the reader has followed all our previous considerations, he will have no further difficulty in understanding the methods leading to the solution of the problem of gravitation.

We start off from a consideration of a Galileian domain, *i.e.* a domain in which there is no gravitational field relative to the Galileian reference-body K. The behaviour of measuring-rods and clocks with reference to K is known from the special theory of relativity, likewise the behaviour of "isolated" material points; the latter move uniformly and in straight lines.

Now let us refer this domain to a random Gauss co-ordinate system or to a "mollusc" as reference-body K'. Then with respect to K' there is a gravitational field G (of a particular kind). We learn the behaviour of measuring-rods and clocks and also of freely-moving material points with reference to K' simply by mathematical transformation. We interpret this behaviour as the behaviour of measuring-rods, clocks and material points under the influence of the gravitational field G. Hereupon we introduce a hypothesis: that the influence of the gravitational field on measuring-rods, clocks and freely-moving material points continues to take place according to the same laws, even in the case where the prevailing gravitational field is *not* derivable from the Galileian special case, simply by means of a transformation of co-ordinates.

The next step is to investigate the space-time behaviour of the gravitational field *G*, which was derived from the Galileian special case simply by transformation of the co-ordinates. This behaviour is formulated in a law, which is always valid, no matter how the reference-body (mollusc) used in the description may be chosen.

This law is not yet the *general* law of the gravitational field, since the gravitational field under consideration is of a special kind. In order to find out the general law-of-field of gravitation we still require to obtain a generalisation of the law as found above. This can be obtained without caprice, however, by taking into consideration the following demands:

(*a*) The required generalisation must likewise satisfy the general postulate of relativity.

(*b*) If there is any matter in the domain under consideration, only its inertial mass, and thus according to Section 15 only its energy is of importance for its effect in exciting a field.

(*c*) Gravitational field and matter together must satisfy the law of the conservation of energy (and of impulse).

Finally, the general principle of relativity permits us to determine the influence of the gravitational field on the course of all those processes which take place according to known laws when a gravitational field is absent, *i.e.* which have already been fitted into the frame of the special theory of relativity. In this connection we proceed in principle according to the method which has already been explained for measuring-rods, clocks and freely-moving material points.

The theory of gravitation derived in this way from the general postulate of relativity excels not only in its beauty; nor in removing the defect attaching to classical mechanics which was brought to light in Section 21; nor in interpreting the empirical law of the equality of inertial and gravitational mass; but it has also already explained a result of observation in astronomy, against which classical mechanics is powerless.

If we confine the application of the theory to the case where the gravitational fields can be regarded as being weak, and in which all masses move with respect to the co-ordinate system with velocities which are small compared with the velocity of light, we then obtain as a first approximation the Newtonian theory. Thus the latter theory is obtained here without any particular assumption, whereas Newton had to introduce the hypothesis that the force of attraction between mutually attracting material points is inversely proportional to the square of the distance between them. If we increase the accuracy of the calculation, deviations from the theory of Newton make their appearance, practically all of which must nevertheless escape the test of observation owing to their smallness.

We must draw attention here to one of these deviations. According to Newton's theory, a planet moves round the sun in an ellipse, which would permanently maintain its position with respect to the fixed stars, if we could disregard the motion of the fixed stars themselves and the action of the other planets under consideration. Thus, if we correct the observed motion of the planets for these two influences, and if Newton's theory be strictly correct, we ought to obtain for the orbit of the planet an ellipse, which is fixed with reference to the fixed stars. This deduction, which can be tested with great accuracy, has been confirmed for all the planets save one, with the precision that is capable of being obtained by the delicacy of observation attainable at the present time. The sole exception is Mercury, the planet which lies nearest the sun. Since the time of Leverrier, it has been known that the ellipse corresponding to the orbit of Mercury, after it has been corrected for the influences mentioned above, is not stationary with respect to the fixed stars, but that it rotates exceedingly slowly in the plane of the orbit and in the sense of the orbital motion. The value obtained for this rotary movement of the orbital ellipse was 43 seconds of arc per century, an amount ensured to be correct to within a few seconds of arc. This effect can be explained by means of classical mechanics only on the assumption of hypotheses which have little probability, and which were devised solely for this purpose.

On the basis of the general theory of relativity, it is found that the ellipse of every planet round the sun must necessarily rotate in the manner indicated above; that for all the planets, with the exception of Mercury, this rotation is too small to be detected with the delicacy of observation possible at the present time; but that in the case of Mercury it must amount to 43 seconds of arc per century, a result which is strictly in agreement with observation.

Apart from this one, it has hitherto been possible to make

only two deductions from the theory which admit of being tested by observation, to wit, the curvature of light rays by the gravitational field of the sun,[1] and a displacement of the spectral lines of light reaching us from large stars, as compared with the corresponding lines for light produced in an analogous manner terrestrially (*i.e.* by the same kind of atom).[2] These two deductions from the theory have both been confirmed.

[1] First observed by Eddington and others in 1919. (Cf. Appendix 3, pp. 145–151.)
[2] Established by Adams in 1924. (Cf. p. 151.)

Part III

Considerations on the Universe as a Whole

THIRTY

Cosmological Difficulties of Newton's Theory

Apart from the difficulty discussed in Section 21, there is a second fundamental difficulty attending classical celestial mechanics, which, to the best of my knowledge, was first discussed in detail by the astronomer Seeliger. If we ponder over the questions as to how the universe, considered as a whole, is to be regarded, the first answer that suggests itself to us is surely this: As regards space (and time) the universe is infinite. There are stars everywhere, so that the density of matter, although very variable in detail, is nevertheless on the average everywhere the same. In other words: However far we might travel through space, we should find everywhere an attenuated swarm of fixed stars of approximately the same kind and density.

This view is not in harmony with the theory of Newton. The latter theory rather requires that the universe should have a kind of centre in which the density of the stars is a maximum, and that as we proceed outwards from this centre the group-density of the stars should diminish, until finally, at great distances, it is succeeded by an infinite region of emptiness. The stellar universe ought to be a finite island in the infinite ocean of space.[1]

This conception is in itself not very satisfactory. It is still less satisfactory because it leads to the result that the light emitted by the stars and also individual stars of the stellar system are perpetually passing out into infinite space, never to return, and without ever again coming into interaction with other objects of nature. Such a finite material universe would be destined to become gradually but systematically impoverished.

In order to escape this dilemma, Seeliger suggested a modification of Newton's law, in which he assumes that for great distances the force of attraction between two masses diminishes more rapidly than would result from the inverse square law. In this way it is possible for the mean density of matter to be constant everywhere, even to infinity, without infinitely large gravitational fields being produced. We thus free ourselves from the distasteful conception that the material universe ought to possess something of the nature of a centre. Of course we purchase our emancipation from the fundamental difficulties mentioned, at the cost of a modification and complication of Newton's law which has neither empirical nor

[1] *Proof*—According to the theory of Newton, the number of "lines of force" which come from infinity and terminate in a mass m is proportional to the mass m. If, on the average, the mass density p_0 is constant throughout the universe, then a sphere of volume V will enclose the average mass p_0V. Thus the number of lines of force passing through the surface F of the sphere into its interior is proportional to p_0V. For unit area of the surface of the sphere the number of lines of force which enters the sphere is thus proportional to $p_0 \frac{V}{F}$ or to p_0R. Hence the intensity of the field at the surface would ultimately become infinite with increasing radius R of the sphere, which is impossible.

theoretical foundation. We can imagine innumerable laws which would serve the same purpose, without our being able to state a reason why one of them is to be preferred to the others; for any one of these laws would be founded just as little on more general theoretical principles as is the law of Newton.

THIRTY-ONE

The Possibility of a "Finite" and Yet "Unbounded" Universe

But speculations on the structure of the universe also move in quite another direction. The development of non-Euclidean geometry led to the recognition of the fact, that we can cast doubt on the *infiniteness* of our space without coming into conflict with the laws of thought or with experience (Riemann, Helmholtz). These questions have already been treated in detail and with unsurpassable lucidity by Helmholtz and Poincaré, whereas I can only touch on them briefly here.

In the first place, we imagine an existence in two-dimensional space. Flat beings with flat implements, and in particular flat rigid measuring-rods, are free to move in a *plane*. For them nothing exists outside of this plane: that which they observe to happen to themselves and to their flat "things" is the all-inclusive reality of their plane. In particular, the constructions of plane Euclidean geometry can be carried out by means of the rods, *e.g.* the lattice construction, considered in Section 24. In contrast to ours, the universe of these beings is two-dimensional; but, like ours, it extends to infinity. In their universe there is room for an infinite number of identical squares made up of rods, *i.e.* its volume (surface) is infinite. If these beings say their universe is "plane," there is sense in the statement, because they mean that they can perform the constructions of plane Euclidean geometry with their rods. In this connection the individual rods always represent the same distance, independently of their position.

Let us consider now a second two-dimensional existence, but this time on a spherical surface instead of on a plane. The flat beings with their measuring-rods and other objects fit exactly on this surface and they are unable to leave it. Their whole universe of observation extends exclusively over the surface of the sphere. Are these beings able to regard the geometry of their universe as being plane geometry and their rods withal as the realisation of "distance"? They cannot do this. For if they attempt to realise a straight line, they will obtain a curve, which we "three-dimensional beings" designate as a great circle, *i.e.* a self-contained line of definite finite length, which can be measured up by means of a measuring-rod. Similarly, this universe has a finite area that can be compared with the area of a square constructed with rods. The great charm resulting from this consideration lies in the recognition of the fact that *the universe of these beings is finite and yet has no limits.*

But the spherical-surface beings do not need to go on a world-tour in order to perceive that they are not living in a Euclidean universe. They can convince themselves of this on every part of their "world," provided they do not use too small a piece of it. Starting from a point, they draw "straight lines" (arcs of circles as judged in three-dimensional space) of equal length in all directions. They will call the line joining the free ends of these lines a "circle." For a plane surface, the ratio of the circumference of a circle to its diameter, both lengths being measured with the same rod, is, according to Euclidean geometry of the plane, equal to a constant value π, which is independent of the diameter of the circle. On their spherical surface our flat beings would find for this ratio the value

$$\pi \frac{\sin\left(\frac{r}{R}\right)}{\left(\frac{r}{R}\right)}$$

i.e. a smaller value than π, the difference being the more considerable, the greater is the radius of the circle in comparison with the radius R of the "world-sphere." By means of this relation the spherical beings can determine the radius of their universe ("world"), even when only a relatively small part of their world-sphere is available for their measurements. But if this part is very small indeed, they will no longer be able to demonstrate that they are on a spherical "world" and not on a Euclidean plane, for a small part of a spherical surface differs only slightly from a piece of a plane of the same size.

Thus if the spherical-surface beings are living on a planet of which the solar system occupies only a negligibly small part of the spherical universe, they have no means of determining whether they are living in a finite or in an infinite universe, because the "piece of universe" to which they have access is in both cases practically plane, or Euclidean. It follows directly from this discussion, that for our sphere-beings the circumference of a circle first increases with the radius until the "circumference of the universe" is reached, and that it thenceforward gradually decreases to zero for still further increasing values of the radius. During this process the area of the circle continues to increase more and more, until finally it becomes equal to the total area of the whole "world-sphere."

Perhaps the reader will wonder why we have placed our "beings" on a sphere rather than on another closed surface. But this choice has its justification in the fact that, of all closed surfaces, the sphere is unique in possessing the property that all points on it are equivalent. I admit that the ratio of the circumference c of circle to its radius r depends on r, but for a given value of r it is the same for all points of the "world-sphere"; in other words, the "world-sphere" is a "surface of constant curvature."

To this two-dimensional sphere-universe there is a three-dimensional analogy, namely, the three-dimensional spherical space which was discovered by Riemann. Its points are likewise all equivalent. It possesses a finite volume, which is determined by its "radius" ($2\pi^2R^3$). Is it possible to imagine a spherical space? To imagine a space means nothing else than that we imagine an epitome of our "space" experience, *i.e.* of experience that we can have in the movement of "rigid" bodies. In this sense we *can* imagine a spherical space.

Suppose we draw lines or stretch strings in all directions

from a point, and mark off from each of these the distance γ with a measuring-rod. All the free end-points of these lengths lie on a spherical surface. We can specially measure up the area (F) of this surface by means of a square made up of measuring-rods. If the universe is Euclidean, then $F = 4\pi\gamma^2$; if it is spherical, then F is always less than $4\pi\gamma^2$. With increasing values of γ, F increases from zero up to a maximum value which is determined by the "world-radius," but for still further increasing values of γ, the area gradually diminishes to zero. At first, the straight lines which radiate from the starting point diverge farther and farther from one another, but later they approach each other, and finally they run together again at a "counter-point" to the starting point. Under such conditions they have traversed the whole spherical space. It is easily seen that the three-dimensional spherical space is quite analogous to the two-dimensional spherical surface. It is finite (*i.e.* of finite volume), and has no bounds.

It may be mentioned that there is yet another kind of curved space: "elliptical space." It can be regarded as a curved space in which the two "counter-points" are identical (indistinguishable from each other). An elliptical universe can thus be considered to some extent as a curved universe possessing central symmetry.

It follows from what has been said, that closed spaces without limits are conceivable. From amongst these, the spherical space (and the elliptical) excels in its simplicity, since all points on it are equivalent. As a result of this discussion, a most interesting question arises for astronomers and physicists, and that is whether the universe in which we live is infinite, or whether it is finite in the manner of the spherical universe. Our experience is far from being sufficient to enable us to answer this question. But the general theory of relativity permits of our answering it with a moderate degree of certainty, and in this connection the difficulty mentioned in Section 30 finds its solution.

THIRTY-TWO

The Structure of Space According to the General Theory of Relativity

According to the general theory of relativity, the geometrical properties of space are not independent, but they are determined by matter. Thus we can draw conclusions about the geometrical structure of the universe only if we base our considerations on the state of the matter as being something that is known. We know from experience that, for a suitably chosen co-ordinate system, the velocities of the stars are small as compared with the velocity of transmission of light. We can thus as a rough approximation arrive at a conclusion as to the nature of the universe as a whole, if we treat the matter as being at rest.

We already know from our previous discussion that the behaviour of measuring-rods and clocks is influenced by gravitational fields, *i.e.* by the distribution of matter. This in itself is sufficient to exclude the possibility of the exact validity of Euclidean geometry in our universe. But it is conceivable that our universe differs only slightly from a Euclidean one, and this notion seems all the more probable, since calculations show that the metrics of surrounding space is influenced only to an exceedingly small extent by masses even of the magnitude of our sun. We might imagine that, as regards geometry, our universe behaves analogously to a surface which is irregularly curved in its individual parts, but which nowhere departs appreciably from a plane: something like the rippled surface of a lake. Such a universe might fittingly be called a quasi-Euclidean universe. As regards its space it would be infinite. But calculation shows that in a quasi-Euclidean universe the average density of matter would necessarily be *nil*. Thus such a universe could not be inhabited by matter everywhere; it would present to us that unsatisfactory picture which we portrayed in Section 30.

If we are to have in the universe an average density of matter which differs from zero, however small may be that difference, then the universe cannot be quasi-Euclidean. On the contrary, the results of calculation indicate that if matter be distributed uniformly, the universe would necessarily be spherical (or elliptical). Since in reality the detailed distribution of matter is not uniform, the real universe will deviate in individual parts from the spherical, *i.e.* the universe will be quasi-spherical. But it will be necessarily finite. In fact, the theory supplies us with a simple connection[1] between the space-expanse of the universe and the average density of matter in it.

[1] For the "radius" R of the universe we obtain the equation

$$R^2 = \frac{2}{\kappa\rho}.$$

The use of the C.G.S. system in this equation gives $\frac{2}{\kappa} = 108.10^{27}$; ρ is the average density of the matter and x is a constant connected with the Newtonian constant of gravitational.

APPENDIX ONE

Simple Derivation of the Lorentz Transformation

[SUPPLEMENTARY TO SECTION 11]

For the relative orientation of the co-ordinate systems indicated in Fig. 2, the x-axes of both systems permanently coincide. In the present case we can divide the problem into parts by considering first only events which are localised on the x-axis. Any such event is represented with respect to the co-ordinate system K by the abscissa x and the time t, and with respect to the system K' by the abscissa x' and the time t'. We require to find x' and t' when x and t are given.

A light-signal, which is proceeding along the positive axis of x, is transmitted according to the equation

$$x = ct$$

or

$$x - ct = 0 \quad . \quad . \quad . \quad (1).$$

Since the same light-signal has to be transmitted relative to K' with the velocity c, the propagation relative to the system K' will be represented by the analogous formula

$$x' - ct' = 0 \quad . \quad . \quad . \quad (2).$$

Those space-time points (events) which satisfy (1) must also satisfy (2). Obviously this will be the case when the relation

$$(x' - ct') = \lambda(x - ct) \quad . \quad . \quad . \quad (3),$$

is fulfilled in general, where λ indicates a constant; for, according to (3), the disappearance of $(x - ct)$ involves the disappearance of $(x' - ct')$.

If we apply quite similar considerations to light rays which are being transmitted along the negative x-axis, we obtain the condition

$$(x' + ct') = \mu(x + ct) \quad . \quad . \quad . \quad (4).$$

By adding (or subtracting) equations (3) and (4), and introducing for convenience the constants a and b in place of the constants λ and μ, where

$$a = \frac{\gamma + \mu}{2}$$

and

$$b = \frac{\gamma - \mu}{2},$$

we obtain the equations

$$\left.\begin{aligned} x' &= ax - bct \\ ct' &= act - bx \end{aligned}\right\} \quad . \quad . \quad . \quad (5).$$

We should thus have the solution of our problem, if the constants a and b were known. These result from the following discussion.

For the origin of K' we have permanently $x' = 0$, and hence according to the first of the equations (5)

$$x = \frac{bc}{a}t.$$

If we call v the velocity with which the origin of K' is moving relative to K, we then have

$$v = \frac{bc}{a} \quad . \quad . \quad . \quad (6).$$

The same value v can be obtained from equations (5), if we calculate the velocity of another point of K' relative to K, or the velocity (directed towards the negative x-axis) of a point of K with respect to K'. In short, we can designate v as the relative velocity of the two systems.

Furthermore, the principle of relativity teaches us that, as judged from K, the length of a unit measuring-rod which is at rest with reference to K' must be exactly the same as the length, as judged from K', of a unit measuring-rod which is at rest relative to K. In order to see how the points of the x'-axis appear as viewed from K, we only require to take a "snapshot" of K' from K; this means that we have to insert a particular value of t (time of K), *e.g.* $t = 0$. For this value of t we then obtain from the first of the equations (5)

$$x' = ax.$$

Two points of the x'-axis which are separated by the distance $\Delta x' = 1$ when measured in the K' system are thus separated in our instantaneous photograph by the distance

$$\Delta x = \frac{1}{a} \quad . \quad . \quad . \quad (7).$$

But if the snapshot be taken from $K'(t' = 0)$, and if we eliminate t from the equations (5), taking into account the expression (6), we obtain

$$x' = a\left(1 - \frac{v^2}{c^2}\right)x.$$

From this we conclude that two points of the x-axis separated by the distance 1 (relative to K) will be represented on our snapshot by the distance

$$\Delta x' = a\left(1 - \frac{v^2}{c^2}\right) \quad . \quad . \quad . \quad (7a).$$

But from what has been said, the two snapshots must be identical; hence Δx in (7) must be equal to $\Delta x'$ in (7*a*), so that we obtain

$$a^2 = \frac{1}{1 - \frac{v^2}{c^2}} \quad . \quad . \quad . \quad (7b).$$

The equations (6) and (7*b*) determine the constants *a* and *b*. By inserting the values of these constants in (5), we obtain the first and the fourth of the equations given in Section XI.

$$\left.\begin{aligned} x' &= \frac{x - vt}{\sqrt{1 - \frac{v^2}{c^2}}} \\ t' &= \frac{t - \frac{v}{c^2}x}{\sqrt{1 - \frac{v^2}{c^2}}} \end{aligned}\right\} \quad . \quad . \quad . \quad (8).$$

Thus we have obtained the Lorentz transformation for events on the *x*-axis. It satisfies the condition

$$x'^2 - c^2t'^2 = x^2 - c^2t^2 \quad . \quad . \quad . \quad (8a).$$

The extension of this result, to include events which take place outside the *x*-axis, is obtained by retaining equations (8) and supplementing them by the relations

$$\left.\begin{aligned} y' &= y \\ z' &= z \end{aligned}\right\} \qquad . \quad . \quad . \quad . \quad (9).$$

In this way we satisfy the postulate of the constancy of the velocity of light *in vacuo* for rays of light of arbitrary directions, both for the system K and for the system K'. This may be shown in the following manner.

We suppose a light-signal sent out from the origin of K at the time $t = 0$. It will be propagated according to the equation

$$r = \sqrt{x^2 + y^2 + z^2} = ct,$$

or, if we square this equation, according to the equation

$$x^2 + y^2 + z^2 - c^2t^2 = 0 \quad . \quad . \quad . \quad (10).$$

It is required by the law of propagation of light, in conjunction with the postulate of relativity, that the transmission of the signal in question should take place—as judged from K'—in accordance with the corresponding formula

$$r' = ct',$$

or,

$$x'^2 + y'^2 + z'^2 - c^2t'^2 = 0 \quad . \quad . \quad (10a).$$

In order that equation (10*a*) may be a consequence of equation (10), we must have

$$x'^2 + y'^2 + z'^2 - c^2t'^2 = \sigma(x^2 + y^2 + z^2 - c^2t^2) \qquad (11).$$

Since equation (8*a*) must hold for points on the x-axis, we thus have $\sigma = 1$. It is easily seen that the Lorentz transformation really satisfies equation (11) for $\sigma = 1$; for (11) is a consequence of (8*a*) and (9), and hence also of (8) and (9). We have thus derived the Lorentz transformation.

The Lorentz transformation represented by (8) and (9) still requires to be generalised. Obviously it is immaterial whether the axes of K' be chosen so that they are spatially parallel to those of K. It is also not essential that the velocity of translation of K' with respect to K should be in the direction of the x-axis. A simple consideration shows that we are able to construct the Lorentz transformation in this general sense from two kinds of transformations, viz. from Lorentz transformations in the special sense and from purely spatial transformations, which corresponds to the replacement of the rectangular co-ordinate system by a new system with its axes pointing in other directions.

Mathematically, we can characterise the generalised Lorentz transformation thus:

It expresses x', y', z', t', in terms of linear homogeneous functions of x, y, z, t, of such a kind that the relation

$$x'^2 + y'^2 + z'^2 - c^2t'^2 = x^2 + y^2 + z^2 - c^2t^2 \qquad (11a)$$

is satisfied identically. That is to say: If we substitute their expressions in x, y, z, t in place of x', y', z', t', on the left-hand side, then the left-hand side of (11*a*) agrees with the right-hand side.

APPENDIX TWO

Minkowski's Four-Dimensional Space ("World")

[SUPPLEMENTARY TO SECTION 17]

We can characterise the Lorentz transformation still more simply if we introduce the imaginary $\sqrt{-1}.\,ct$ in place of t, as time-variable. If, in accordance with this, we insert

$$\begin{aligned} x_1 &= x \\ x_2 &= y \\ x_3 &= z \\ x_4 &= \sqrt{-1}.\,ct, \end{aligned}$$

and similarly for the accented system K', then the condition which is identically satisfied by the transformation can be expressed thus:

$$x_1'^2 + x_2'^2 + x_3'^2 + x_4'^2 = x_1^2 + x_2^2 + x_3^2 + x_4^2 \qquad (12).$$

That is, by the afore-mentioned choice of "co-ordinates," (11*a*) is transformed into this equation.

We see from (12) that the imaginary time co-ordinate x_4 enters into the condition of transformation in exactly the same way as the space co-ordinates x_1, x_2, x_3. It is due to this fact that, according to the theory of relativity, the "time" x_4 enters into natural laws in the same form as the space co-ordinates x_1, x_2, x_3.

A four-dimensional continuum described by the "co-ordinates" x_1, x_2, x_3, x_4, was called "world" by Minkowski, who also termed a point-event a "world-point." From a "happening" in three-dimensional space, physics becomes, as it were, an "existence" in the four-dimensional "world."

This four-dimensional "world" bears a close similarity to the three-dimensional "space" of (Euclidean) analytical geometry. If we introduce into the latter a new Cartesian co-ordinate system (x'_1, x'_2, x'_3) with the same origin, then x'_1, x'_2, x'_3, are linear homogeneous functions of x_1, x_2, x_3, which identically satisfy the equation

$$x_1'^2 + x_2'^2 + x_3'^2 = x_1^2 + x_2^2 + x_3^2.$$

The analogy with (12) is a complete one. We can regard Minkowski's "world" in a formal manner as a four-dimensional Euclidean space (with imaginary time co-ordinate); the Lorentz transformation corresponds to a "rotation" of the co-ordinate system in the four-dimensional "world."

APPENDIX THREE

The Experimental Confirmation of the General Theory of Relativity

From a systematic theoretical point of view, we may imagine the process of evolution of an empirical science to be a continuous process of induction. Theories are evolved and are expressed in short compass as statements of a large number of individual observations in the form of empirical laws, from which the general laws can be ascertained by comparison. Regarded in this way, the development of a science bears some resemblance to the compilation of a classified catalogue. It is, as it were, a purely empirical enterprise.

But this point of view by no means embraces the whole of the actual process; for it slurs over the important part played by intuition and deductive thought in the development of an exact science. As soon as a science has emerged from its initial stages, theoretical advances are no longer achieved merely by a process of arrangement. Guided by empirical data, the investigator rather develops a system of thought which, in general, is built up logically from a small number of fundamental assumptions, the so-called axioms. We call such a system of thought a *theory*. The theory finds the justification for its existence in the fact that it correlates a large number of single observations, and it is just here that the "truth" of the theory lies.

Corresponding to the same complex of empirical data, there may be several theories, which differ from one another to a considerable extent. But as regards the deductions from the theories which are capable of being tested, the agreement between the theories may be so complete, that it becomes difficult to find any deductions in which the two theories differ from each other. As an example, a case of general interest is available in the province of biology, in the Darwinian theory of the development of species by selection in the struggle for existence, and in the theory of development which is based on the hypothesis of the hereditary transmission of acquired characters.

We have another instance of far-reaching agreement between the deductions from two theories in Newtonian mechanics on the one hand, and the general theory of relativity on the other. This agreement goes so far, that up to the present we have been able to find only a few deductions from the general theory of relativity which are capable of investigation, and to which the physics of pre-relativity days does not also lead, and this despite the profound difference in the fundamental assumptions of the two theories. In what follows, we shall again consider these important deductions, and we shall also discuss the empirical evidence appertaining to them which has hitherto been obtained.

(a) Motion of the Perihelion of Mercury

According to Newtonian mechanics and Newton's law of gravitation, a planet which is revolving round the sun would describe an ellipse round the latter, or, more correctly, round the common centre of gravity of the sun and the planet. In such a system, the sun, or the common centre of gravity, lies in one of the foci of the orbital ellipse in such a manner that, in the course of a planet-year, the distance sun-planet grows from a minimum to a maximum, and then decreases again to a minimum. If instead of Newton's law we insert a somewhat different law of attraction into the calculation, we find that, according to this new law, the motion would still take place in such a manner that the distance sun-planet exhibits periodic variations; but in this case the angle described by the line joining sun and planet during such a period (from perihelion—closest proximity to the sun—to perihelion) would differ from 360°. The line of the orbit would not then be a closed one but in the course of time it would fill up an annular part of the orbital plane, viz. between the circle of least and the circle of greatest distance of the planet from the sun.

According also to the general theory of relativity, which differs of course from the theory of Newton, a small variation from the Newton-Kepler motion of a planet in its orbit should take place, and in such a way, that the angle described by the radius sun-planet between one perihelion and the next should exceed that corresponding to one complete revolution by an amount given by

$$+ \frac{24\pi^3 a^2}{T^2 c^2 (1 - e^2)}.$$

(*N.B.*—One complete revolution corresponds to the angle 2π in the absolute angular measure customary in physics, and the above expression gives the amount by which the radius sun-planet exceeds this angle during the interval between one perihelion and the next.) In this expression a represents the major semi-axis of the ellipse, e its eccentricity, c the velocity of light, and T the period of revolution of the planet. Our result may also be stated as follows: According to the general theory of relativity, the major axis of the ellipse rotates round the sun in the same sense as the orbital motion of the planet. Theory requires that this rotation should amount to 43 seconds of arc per century for the planet Mercury, but for the other planets of our solar system its magnitude should be so small that it would necessarily escape detection.[1]

In point of fact, astronomers have found that the theory of Newton does not suffice to calculate the observed motion of Mercury with an exactness corresponding to that of the delicacy of observation attainable at the present time. After taking account of all the disturbing influences exerted on Mercury by the remaining planets, it was found (Leverrier—1859—and Newcomb—1895) that an unexplained perihelial movement of the orbit of Mercury remained over, the amount of which does not differ sensibly from the above-mentioned +43 sec-

[1] Especially since the next planet Venus has an orbit that is almost an exact circle, which makes it more difficult to locate the perihelion with precision.

onds of arc per century. The uncertainty of the empirical result amounts to a few seconds only.

(B) Deflection of Light by a Gravitational Field

In Section 12 it has been already mentioned that according to the general theory of relativity, a ray of light will experience a curvature of its path when passing through a gravitational field, this curvature being similar to that experienced by the path of a body which is projected through a gravitational field. As a result of this theory, we should expect that a ray of light which is passing close to a heavenly body would be deviated towards the latter. For a ray of light which passes the sun at a distance of Δ sun-radii from its centre, the angle of deflection (a) should amount to

$$a = \frac{1.7 \text{ seconds of arc}}{\Delta}.$$

It may be added that, according to the theory, half of this deflection is produced by the Newtonian field of attraction of the sun, and the other half by the geometrical modification ("curvature") of space caused by the sun.

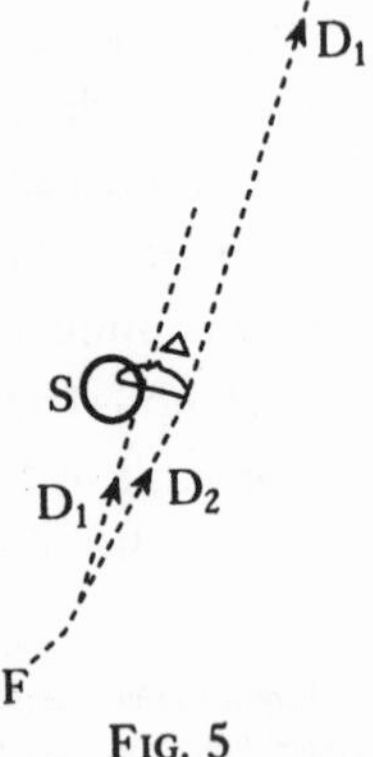

Fig. 5

This result admits of an experimental test by means of the photographic registration of stars during a total eclipse of the sun. The only reason why we must wait for a total eclipse is because at every other time the atmosphere is so strongly illuminated by the light from the sun that the stars situated near the sun's disc are invisible. The predicted effect can be seen clearly from the accompanying diagram. If the sun (S) were not present, a star which is practically infinitely distant would be seen in the direction D_1, as observed from the earth. But as a consequence of the deflection of light from the star by the sun, the star will be seen in the direction D_2, *i.e.* at a somewhat greater distance from the centre of the sun that corresponds to its real position.

In practice, the question is tested in the following way. The stars in the neighbourhood of the sun are photographed during a solar eclipse. In addition, a second photograph of the same stars is taken when the sun is situated at another position in the sky, *i.e.* a few months earlier or later. As compared with the standard photograph, the positions of the stars on the eclipse-photograph ought to appear displaced radially outwards (away from the centre of the sun) by an amount corresponding to the angle a.

We are indebted to the Royal Society and to the Royal Astronomical Society for the investigation of this important deduction. Undaunted by the war and by difficulties of both a material and a psychological nature aroused by the war, these societies equipped two expeditions—to Sobral (Brazil), and to the island of Principe (West Africa)—and sent several of Britain's most celebrated astronomers (Eddington, Cottingham, Crommelin, Davidson), in order to obtain photographs of the solar eclipse of 29th May, 1919. The relative discrepancies to be expected between the stellar photographs obtained during the eclipse and the comparison photographs amounted to a few hundredths of a millimetre only. Thus great accuracy was necessary in making the adjustments required for the taking of the photographs, and in their subsequent measurement.

The results of the measurements confirmed the theory in a thoroughly satisfactory manner. The rectangular components of the observed and of the calculated deviations of the stars (in seconds of arc) are set forth in the following table of results:

Number of the Star.	First Co-ordinate.		Second Co-ordinate.	
	Observed.	Calculated.	Observed.	Calculated.
11	−0.19	−0.22	+0.16	+0.02
5	−0.29	−0.31	−0.46	−0.43
4	−0.11	−0.10	+0.83	+0.74
3	−0.20	−0.12	+1.00	+0.87
6	−0.10	−0.04	+0.57	+0.40
10	−0.08	+0.09	+0.35	+0.32
2	+0.95	+0.85	−0.27	−0.09

(c) Displacement of Spectral Lines towards the Red

In Section 23 it has been shown that in a system K' which is in rotation with regard to a Galileian system K, clocks of identical construction, and which are considered at rest with respect to the rotating reference-body, go at rates which are dependent on the positions of the clocks. We shall now examine this dependence quantitatively. A clock, which is situated at a distance r from the centre of the disc, has a velocity relative to K which is given by

$$v = \omega r$$

where ω represents the angular velocity of rotation of the disc K' with respect to K. If ν_0 represents the number of ticks of the clock per unit time ("rate" of the clock) relative to K when the clock is at rest, then the "rate" of the clock (ν) when it is moving relative to K with a velocity v, but at rest with respect to the disc, will, in accordance with Section 12, be given by

$$\nu = \nu_0 \sqrt{1 - \frac{v^2}{c^2}},$$

or with sufficient accuracy by

$$\nu = \nu_0 \left(1 - \frac{1}{2} \frac{v^2}{c^2}\right).$$

This expression may also be stated in the following form:

$$\nu = \nu_0 \left(1 - \frac{1}{c^2} \frac{\omega^2 r^2}{2}\right).$$

If we represent the difference of potential of the centrifugal force between the position of the clock and the centre of the disc by ϕ, *i.e.* the work, considered negatively, which must be performed on the unit of mass against the centrifugal force in order to transport it from the position of the clock on the rotating disc to the centre of the disc, then we have

$$\phi = -\frac{\omega^2 r^2}{2}.$$

From this it follows that

$$\nu = \nu_0\left(1 + \frac{\phi}{c^2}\right).$$

In the first place, we see from this expression that two clocks of identical construction will go at different rates when situated at different distances from the centre of the disc. This result is also valid from the standpoint of an observer who is rotating with the disc.

Now, as judged from the disc, the latter is in a gravitational field of potential ϕ, hence the result we have obtained will hold quite generally for gravitational fields. Furthermore, we can regard an atom which is emitting spectral lines as a clock, so that the following statement will hold:

An atom absorbs or emits light of a frequency which is dependent on the potential of the gravitational field in which it is situated.

The frequency of an atom situated on the surface of a heavenly body will be somewhat less than the frequency of an atom of the same element which is situated in free space (or on the surface of a smaller celestial body). Now $\phi = -K\frac{M}{r}$, where K is Newton's constant of gravitation, and M is the mass of the heavenly body. Thus a displacement towards the red ought to take place for spectral lines produced at the surface of stars as compared with the spectral lines of the same element produced at the surface of the earth, the amount of this displacement being

$$\frac{\nu - \nu_0}{\nu} = \frac{K}{c^2}\frac{M}{r}.$$

For the sun, the displacement towards the red predicted by theory amounts to about two millionths of the wave-length. A trustworthy calculation is not possible in the case of the stars, because in general neither the mass M nor the radius r are known.

It is an open question whether or not this effect exists, and at the present time (1920) astronomers are working with great zeal towards the solution. Owing to the smallness of the effect in the case of the sun, it is difficult to form an opinion as to its existence. Whereas Grebe and Bachem (Bonn), as a result of their own measurements and those of Evershed and Schwarzschild on the cyanogen bands, have placed the existence of the effect almost beyond doubt, other investigators, particularly St. John, have been led to the opposite opinion in consequence of their measurements.

Mean displacements of lines towards the less refrangible end of the spectrum are certainly revealed by statistical investigations of the fixed stars; but up to the present the examination of the available data does not allow of any definite decision being arrived at, as to whether or not these displacements are to be referred in reality to the effect of gravitation. The results of observation have been collected together, and discussed in detail from the standpoint of the question which has been engaging our attention here, in a paper by E. Freundlich entitled "Zur Prüfung der aligemeinen Relativitäts-Theorie" (*Die Naturwissenschaften*, 1919, No. 35, p. 520: Julius Springer, Berlin).

At all events, a definite decision will be reached during the next few years. If the displacement of spectral lines towards the red by the gravitational potential does not exist, then the general theory of relativity will be untenable. On the other hand, if the cause of the displacement of spectral lines be definitely traced to the gravitational potential, then the study of this displacement will furnish us with important information as to the mass of the heavenly bodies.

NOTE.—The displacement of spectral lines towards the red end of the spectrum was definitely established by Adams in 1924, by observations on the dense companion of Sirius, for which the effect is about thirty times greater than for the sun.

R. W. L.

APPENDIX FOUR

The Structure of Space According to the General Theory of Relativity

[SUPPLEMENTARY TO SECTION 32]

Since the publication of the first edition of this little book, our knowledge about the structure of space in the large ("cosmological problem") has had an important development, which ought to be mentioned even in a popular presentation of the subject.

My original considerations on the subject were based on two hypotheses:

1. There exists an average density of matter in the whole of space which is everywhere the same and different from zero.
2. The magnitude ("radius") of space is independent of time.

Both these hypotheses proved to be consistent, according to the general theory of relativity, but only after a hypothetical term was added to the field equations, a term which was not required by the theory as such nor did it seem natural from a theoretical point of view ("cosmological term of the field equations").

Hypothesis (2) appeared unavoidable to me at the time, since I thought that one would get into bottomless speculations if one departed from it.

However, already in the 'twenties, the Russian mathematician Friedman showed that a different hypothesis was natural from a purely theoretical point of view. He realized that it was possible to preserve hypothesis (1) without introducing the less natural cosmological term into the field equations of gravitation, if one was ready to drop hypothesis (2). Namely, the original field equations admit a solution in which the "world-radius" depends on time (expanding space). In that sense one can say, according to Friedman, that the theory demands an expansion of space.

A few years later Hubble showed, by a special investigation of the extra-galactic nebulae ("milky ways"), that the spectral lines emitted showed a red shift which increased regularly with the distance of the nebulae. This can be interpreted in regard to our present knowledge only in the sense of Doppler's principle, as an expansive motion of the system of stars in the large—as required, according to Friedman, by the field equations of gravitation. Hubble's discovery can, therefore, be considered to some extent as a confirmation of the theory.

There does arise, however, a strange difficulty. The interpretation of the galactic line-shift discovered by Hubble as an expansion (which can hardly be doubted from a theoretical point of view), leads to an origin of this expansion which lies "only" about 10^9 years ago, while physical astronomy makes it appear likely that the development of individual stars and systems of stars takes considerably longer. It is in no way known how this incongruity is to be overcome.

I further want to remark that the theory of expanding space, together with the empirical data of astronomy, permit no decision to be reached about the finite or infinite character of (three-dimensional) space, while the original "static" hypothesis of space yielded the closure (finiteness) of space.

APPENDIX FIVE

Relativity and the Problem of Space[1]

It is characteristic of Newtonian physics that it has to ascribe independent and real existence to space and time as well as to matter, for in Newton's law of motion the idea of acceleration appears. But in this theory, acceleration can only denote "acceleration with respect to space." Newton's space must thus be thought of as "at rest," or at least as "unaccelerated," in order that one can consider the acceleration, which appears in the law of motion, as being a magnitude with any meaning. Much the same holds with time, which of course likewise enters into the concept of acceleration. Newton himself and his most critical contemporaries felt it to be disturbing that one had to ascribe physical reality both to space itself as well as to its state of motion; but there was at that time no other alternative, if one wished to ascribe to mechanics a clear meaning.

[1] As with the original translation of this book in 1920, my old friend Emeritus Professor S. R. Milner, F.R.S. has again given me the benefit of his unique experience in this field, by reading the translation of this new appendix and making numerous suggestions for improvement. I am deeply grateful to him and to Professor A. G. Walker of the Mathematics Department of Liverpool University, who also read this appendix and offered various helpful suggestions.

R. W. L.

It is indeed an exacting requirement to have to ascribe physical reality to space in general, and especially to empty space. Time and again since remotest times philosophers have resisted such a presumption. Descartes argued somewhat on these lines: space is identical with extension, but extension is connected with bodies; thus there is no space without bodies and hence no empty space. The weakness of this argument lies primarily in what follows. It is certainly true that the concept extension owes its origin to our experiences of laying out or bringing into contact solid bodies. But from this it cannot be concluded that the concept of extension may not be justified in cases which have not themselves given rise to the formation of this concept. Such an enlargement of concepts can be justified indirectly by its value for the comprehension of empirical results. The assertion that extension is confined to bodies is therefore of itself certainly unfounded. We shall see later, however, that the general theory of relativity confirms Descartes' conception in a roundabout way. What brought Descartes to his remarkably attractive view was certainly the feeling that, without compelling necessity, one ought not to ascribe reality to a thing like space, which is not capable of being "directly experienced."[1]

The psychological origin of the idea of space, or of the necessity for it, is far from being so obvious as it may appear to be on the basis of our customary habit of thought. The old geometers deal with conceptual objects (straight line, point, surface), but not really with space as such, as was done later in

[1] This expression is to be taken *cum grano salis.*

analytical geometry. The idea of space, however, is suggested by certain primitive experiences. Suppose that a box has been constructed. Objects can be arranged in a certain way inside the box, so that it becomes full. The possibility of such arrangements is a property of the material object "box," something that is given with the box, the "space enclosed" by the box. This is something which is different for different boxes, something that is thought quite naturally at being independent of whether or not, at any moment, there are any objects at all in the box. When there are no objects in the box, its space appears to be "empty."

So far, our concept of space has been associated with the box. It turns out, however, that the storage possibilities that make up this box-space are independent of the thickness of the walls of the box. Cannot this thickness be reduced to zero, without the "space" being lost as a result? The naturalness of such a limiting process is obvious, and now there remains for our thought the space without the box, a self-evident thing, yet it appears to be so unreal if we forget the origin of this concept. One can understand that it was repugnant to Descartes to consider space as independent of material objects, a thing that might exist without matter.[1] (At the same time, this does not prevent him from treating space as a fundamental concept in his analytical geometry.) The drawing of attention to the vacuum in a mercury barometer has certainly disarmed

[1] Kant's attempt to remove the embarrassment by denial of the objectivity of space can, however, hardly be taken seriously. The possibilities of packing inherent in the inside space of a box are objective in the same sense as the box itself, and as the objects which can be packed inside it.

the last of the Cartesians. But it is not to be denied that, even at this primitive stage, something unsatisfactory clings to the concept of space, or to space thought of as an independent real thing.

The ways in which bodies can be packed into space (*e.g.* the box) are the subject of three-dimensional Euclidean geometry, whose axiomatic structure readily deceives us into forgetting that it refers to realisable situations.

If now the concept of space is formed in the manner outlined above, and following on from experience about the "filling" of the box, then this space is primarily a ***bounded*** space. This limitation does not appear to be essential, however, for apparently a larger box can always be introduced to enclose the smaller one. In this way space appears as something unbounded.

I shall not consider here how the concepts of the three-dimensional and the Euclidean nature of space can be traced back to relatively primitive experiences. Rather, I shall consider first of all from other points of view the rôle of the concept of space in the development of physical thought.

When a smaller box s is situated, relatively at rest, inside the hollow space of a larger box S, then the hollow space of s is a part of the hollow space of S, and the same "space," which contains both of them, belongs to each of the boxes. When s is in motion with respect to S, however, the concept is less simple. One is then inclined to think that s encloses always the same space, but a variable part of the space S. It then becomes necessary to apportion to each box its particular space, not thought of as bounded, and to assume that these two spaces are in motion with respect to each other.

Before one has become aware of this complication, space appears as an unbounded medium or container in which material objects swim around. But it must now be remembered that there is an infinite number of spaces, which are in motion with respect to each other. The concept of space as something existing objectively and independent of things belongs to prescientific thought, but not so the idea of the existence of an infinite number of spaces in motion relatively to each other. This latter idea is indeed logically unavoidable, but is far from having played a considerable rôle even in scientific thought.

But what about the psychological origin of the concept of time? This concept is undoubtedly associated with the fact of "calling to mind," as well as with the differentiation between sense experiences and the recollection of these. Of itself it is doubtful whether the differentiation between sense experience and recollection (or simple re-presentation) is something psychologically directly given to us. Everyone has experienced that he has been in doubt whether he has actually experienced something with his senses or has simply dreamt about it. Probably the ability to discriminate between these alternatives first comes about as the result of an activity of the mind creating order.

An experience is associated with a "recollection," and it is considered as being "earlier" in comparison with "present experiences." This is a conceptual ordering principle for recollected experiences, and the possibility of its accomplishment gives rise to the subjective concept of time, *i.e.* that concept of time which refers to the arrangement of the experiences of the individual.

What do we mean by rendering objective the concept of time? Let us consider an example. A person *A* ("I") has the experience "it is lightning." At the same time the person *A* also experiences such a behaviour of the person *B* as brings the behaviour of *B* into relation with his own experience "it is lightning." Thus it comes about that *A* associates with *B* the experience "it is lightning." For the person *A* the idea arises that other persons also participate in the experience "it is lightning." "It is lightning" is now no longer interpreted as an exclusively personal experience, but as an experience of other persons (or eventually only as a "potential experience"). In this way arises the interpretation that "it is lightning," which originally entered into the consciousness of an "experience," is now also interpreted as an (objective) "event." It is just the sum total of all events that we mean when we speak of the "real external world."

We have seen that we feel ourselves impelled to ascribe a temporal arrangement to our experiences, somewhat as follows. If β is later than α and γ later than β, then γ is also later than α ("sequence of experiences"). Now what is the position in this respect with the "events" which we have associated with the experiences? At first sight it seems obvious to assume that a temporal arrangement of events exists which agrees with the temporal arrangement of the experiences. In general, and unconsciously this was done, until sceptical doubts made themselves felt.[1] In order to arrive at the idea of an objective world, an additional constructive concept still is necessary: the event is localised not only in time, but also in space.

In the previous paragraphs we have attempted to describe how the concepts space, time and event can be put psychologically into relation with experiences. Considered logically, they are free creations of the human intelligence, tools of thought, which are to serve the purpose of bringing experiences into relation with each other, so that in this way they can be better surveyed. The attempt to become conscious of the empirical sources of these fundamental concepts should show to what extent we are actually bound to these concepts. In this way we become aware of our freedom, of which, in case of necessity, it is always a difficult matter to make sensible use.

We still have something essential to add to this stretch concerning the psychological origin of the concepts space-time-event (we will call them more briefly "space-like," in contrast to concepts from the psychological sphere). We have linked up the concept of space with experiences using boxes and the arrangement of material objects in them. Thus this formation of concepts already presupposes the concept of material objects (*e.g.* "boxes"). In the same way persons, who had to be introduced for the formation of an objective concept of time, also play the rôle of material objects in this connection. It appears to me, therefore, that the formation of the concept

[1] For example, the order of experiences in time obtained by acoustical means can differ from the temporal order gained visually, so that one cannot simply identify the time sequence of events with the time sequence of experiences.

of the material object must precede our concepts of time and space.

All these space-like concepts already belong to pre-scientific thought, along with concepts like pain, goal, purpose, etc. from the field of psychology. Now it is characteristic of thought in physics, as of thought in natural science generally, that it endeavours in principle to make do with "space-like" concepts *alone*, and strives to express with their aid all relations having the form of laws. The physicist seeks to reduce colours and tones to vibrations, the physiologist thought and pain to nerve processes, in such a way that the psychical element as such is eliminated from the causal nexus of existence, and thus nowhere occurs as an independent link in the causal associations. It is no doubt this attitude, which considers the comprehension of all relations by the exclusive use of only "space-like" concepts as being possible in principle, that is at the present time understood by the term "materialism" (since "matter" has lost its rôle as a fundamental concept).

Why is it necessary to drag down from the Olympian fields of Plato the fundamental ideas of thought in natural science, and to attempt to reveal their earthly lineage? Answer: In order to free these ideas from the taboo attached to them, and thus to achieve greater freedom in the formation of ideas or concepts. It is to the immortal credit of D. Hume and E. Mach that they, above all others, introduced this critical conception.

Science has taken over from pre-scientific thought the concepts space, time, and material object (with the important special case "solid body"), and has modified them and rendered them more precise. Its first significant accomplishment was the development of Euclidean geometry, whose axiomatic formulation must not be allowed to blind us to its empirical origin (the possibilities of laying out or juxtaposing solid bodies). In particular, the three-dimensional nature of space as well as its Euclidean character are of empirical origin (it can be wholly filled by like constituted "cubes").

The subtlety of the concept of space was enhanced by the discovery that there exist no completely rigid bodies. All bodies are elastically deformable and alter in volume with change in temperature. The structures, whose possible congruences are to be described by Euclidean geometry, cannot therefore be represented apart from physical concepts. But since physics after all must make use of geometry in the establishment of its concepts, the empirical content of geometry can be stated and tested only in the framework of the whole of physics.

In this connection atomistics must also be borne in mind, and its conception of finite divisibility; for spaces of subatomic extension cannot be measured up. Atomistics also compels us to give up, in principle, the idea of sharply and statically defined bounding surfaces of solid bodies. Strictly speaking, there are no *precise* laws, even in the macro-region, for the possible configurations of solid bodies touching each other.

In spite of this, no one thought of giving up the concept of space, for it appeared indispensable in the eminently satisfactory whole system of natural science. Mach, in the nineteenth century, was the only one who thought seriously of an elimination of the concept of space, in that he sought to replace it by the notion of the totality of the instantaneous distances between all material points. (He made this attempt in order to arrive at a satisfactory understanding of inertia).

The Field

In Newtonian mechanics, space and time play a dual rôle. First, they play the part of carrier or frame for things that happen in physics, in reference to which events are described by the space co-ordinates and the time. In principle, matter is thought of as consisting of "material points," the motions of which constitute physical happening. When matter is thought of as being continuous, this is done as it were provisionally in those cases where one does not wish to or cannot describe the discrete structure. In this case small parts (elements of volume) of the matter are treated similarly to material points, at least in so far as we are concerned merely with motions and not with occurrences which, at the moment, it is not possible or serves no useful purpose to attribute to motion (*e.g.* temperature changes, chemical processes). The second rôle of space and time was that of being an "inertial system." From all conceivable systems of reference, inertial systems were considered to be advantageous in that, with respect to them, the law of inertia claimed validity.

In this, the essential thing is that "physical reality," thought of as being independent of the subjects experiencing it, was conceived as consisting, at least in principle, of space and time on one hand, and of permanently existing material points, moving with respect to space and time, on the other. The idea of the independent existence of space and time can be expressed drastically in this way: If matter were to disappear, space and time alone would remain behind (as a kind of stage for physical happening).

The surmounting of this standpoint resulted from a development which, in the first place, appeared to have nothing to do with the problem of space-time, namely, the appearance of the *concept of field* and its final claim to replace, in principle, the idea of a particle (material point). In the framework of classical physics, the concept of field appeared as an auxiliary concept, in cases in which matter was treated as a continuum. For example, in the consideration of the heat conduction in a solid body, the state of the body is described by giving the temperature at every point of the body for every definite time. Mathematically, this means that the temperature T is represented as a mathematical expression (function) of the space co-ordinates and the time t (Temperature field). The law of heat conduction is represented as a local relation (differential equation), which embraces all special cases of the conduction of heat. The temperature is here a simple example of the concept of field. This is a quantity (or a complex of quantities), which is a function of the co-ordinates and the time. Another example is the description of the motion of a liquid. At every point there exists at any time a velocity, which is quantitatively described by its three "components" with respect to the axes of a co-ordinate system (vector). The components of the velocity at a point (field components), here also, are functions of the co-ordinates (x, y, z) and the time (t).

It is characteristic of the fields mentioned that they occur only within a ponderable mass; they serve only to describe a state of this matter. In accordance with the historical development of the field concept, where no matter was available there could also exist no field. But in the first quarter of the nineteenth century it was shown that the phenomena of the interference and motion of light could be explained with astonishing clearness when light was regarded as a wave-field, completely analogous to the mechanical vibration field in an elastic solid body. It was thus felt necessary to introduce a field, that could also exist in "empty space" in the absence of ponderable matter.

This state of affairs created a paradoxical situation, because, in accordance with its origin, the field concept appeared to be restricted to the description of states in the inside of a ponderable body. This seemed to be all the more certain, inasmuch as the conviction was held that every field is to be regarded as a state capable of mechanical interpretation, and this presupposed the presence of matter. One thus felt compelled, even in the space which had hitherto been regarded as empty, to assume everywhere the existence of a form of matter, which was called "æther."

The emancipation of the field concept from the assumption of its association with a mechanical carrier finds a place among the psychologically most interesting events in the development of physical thought. During the second half of the nineteenth century, in connection with the researches of Faraday and Maxwell, it became more and more clear that the description of electromagnetic processes in terms of field was vastly superior to a treatment on the basis of the mechanical concepts of material points. By the introduction of the field concept in electrodynamics, Maxwell succeeded in predicting the existence of electromagnetic waves, the essential identity of which with light waves could not be doubted, because of the equality of their velocity of propagation. As a result of this, optics was, in principle, absorbed by electrodynamics. *One* psychological effect of this immense success was that the field concept, as opposed to the mechanistic framework of classical physics, gradually won greater independence.

Nevertheless, it was at first taken for granted that electromagnetic fields had to be interpreted as states of the æther, and it was zealously sought to explain these states as mechanical ones. But as these efforts always met with frustration, science gradually became accustomed to the idea of renouncing such a mechanical interpretation. Nevertheless, the conviction still remained that electromagnetic fields must be states of the æther, and this was the position at the turn of the century.

The æther-theory brought with it the question: How does the æther behave from the mechanical point of view with respect to ponderable bodies? Does it take part in the motions of the bodies, or do its parts remain at rest relatively to each other? Many ingenious experiments were undertaken to decide this question. The following important facts should be mentioned in this connection: the "aberration" of the fixed stars in consequence of the annual motion of the earth, and the "Doppler effect," *i.e.* the influence of the relative motion of the fixed stars on the frequency of the light reaching us from them, for known frequencies of emission. The results of all these facts and experiments, except for one, the Michelson-

Morley experiment, were explained by H. A. Lorentz on the assumption that the æther does not take part in the motions of ponderable bodies, and that the parts of the æther have no relative motions at all with respect to each other. Thus the æther appeared, as it were, as the embodiment of a space absolutely at rest. But the investigation of Lorentz accomplished still more. It explained all the electromagnetic and optical processes within ponderable bodies known at that time, on the assumption that the influence of ponderable matter on the electric field—and conversely—is due solely to the fact that the constituent particles of matter carry electrical charges, which share the motion of the particles. Concerning the experiment of Michelson and Morley, H. A. Lorentz showed that the result obtained at least does not contradict the theory of an æther at rest.

In spite of all these beautiful successes the state of the theory was not yet wholly satisfactory, and for the following reasons. Classical mechanics, of which it could not be doubted that it holds with a close degree of approximation, teaches the equivalence of all inertial systems or inertial "spaces" for the formulation of natural laws, *i.e.* the invariance of natural laws with respect to the transition from one inertial system to another. Electromagnetic and optical *experiments* taught the same thing with considerable accuracy. But the foundation of electromagnetic *theory* taught that a particular inertial system must be given preference, namely that of the luminiferous æther at rest. This view of the theoretical foundation was much too unsatisfactory. Was there no modification that, like classical mechanics, would uphold the equivalence of inertial systems (special principle of relativity)?

The answer to this question is the special theory of relativity. This takes over from the theory of Maxwell–Lorentz the assumption of the constancy of the velocity of light in empty space. In order to bring this into harmony with the equivalence of inertial systems (special principle of relativity), the idea of the absolute character of simultaneity must be given up; in addition, the Lorentz transformations for the time and the space co-ordinates follow for the transition from one inertial system to another. The whole content of the special theory of relativity is included in the postulate: The laws of Nature are invariant with respect to the Lorentz transformations. The important thing of this requirement lies in the fact that it limits the possible natural laws in a definite manner.

What is the position of the special theory of relativity in regard to the problem of space? In the first place we must guard against the opinion that the four-dimensionality of reality has been newly introduced for the first time by this theory. Even in classical physics the event is localised by four numbers, three spatial co-ordinates and a time co-ordinate; the totality of physical "events" is thus thought of as being embedded in a four-dimensional continuous manifold. But the basis of classical mechanics this four-dimensional continuum breaks up objectively into the one-dimensional time and into three-dimensional spatial sections, only the latter of which contain simultaneous events. This resolution is the same for all inertial systems. The simultaneity of two definite events with reference to one inertial system involves the simultaneity of these events in reference to all inertial systems. This is what is meant when we say that the time of classical mechanics is absolute. According to the special theory of relativity it is otherwise. The sum total of events which are simultaneous with a selected event exist, it is true, in relation to a particular inertial system, but no longer independently of the choice of the inertial system. The four-dimensional continuum is now no longer resolvable objectively into sections, all of which contain simultaneous events; "now" loses for the spatially extended world its objective meaning. It is because of this that space and time must be regarded as a four-dimensional continuum that is objectively unresolvable, if it is desired to express the purport of objective relations without unnecessary conventional arbitrariness.

Since the special theory of relativity revealed the physical equivalence of all inertial systems, it proved the untenability of the hypothesis of an æther at rest. It was therefore necessary to renounce the idea that the electromagnetic field is to be regarded as a state of a material carrier. The field thus becomes an irreducible element of physical description, irreducible in the same sense as the concept of matter in the theory of Newton.

Up to now we have directed our attention to finding in what respect the concepts of space and time were ***modified*** by the special theory of relativity. Let us now focus our attention on those elements which this theory has taken over from classical mechanics. Here also, natural laws claim validity only when an inertial system is taken as the basis of space-time description. The principle of inertia and the principle of the constancy of the velocity of light are valid only with respect to an *inertial system*. The field-laws also can claim to have a meaning and validity only in regard to inertial systems. Thus, as in classical mechanics, space is here also an independent component in the representation of physical reality. If we imagine matter and field to be removed, inertial-space or, more accurately, this space together with the associated time remains behind. The four-dimensional structure (Minkowski-space) is thought of as being the carrier of matter and of the field. Inertial spaces, with their associated times, are only privileged four-dimensional co-ordinate systems, that are linked together by the linear Lorentz transformations. Since there exist in this four-dimensional structure no longer any sections which represent "now" objectively, the concepts of happening and becoming are indeed not completely suspended, but yet complicated. It appears therefore more natural to think of physical reality as a four-dimensional existence, instead of, as hitherto, the *evolution* of a three-dimensional existence.

This rigid four-dimensional space of the special theory of relativity is to some extent a four-dimensional analogue of H. A. Lorentz's rigid three-dimensional æther. For this theory also the following statement is valid: The description of physical states postulates space as being initially given and as existing independently. Thus even this theory does not dispel Descartes' uneasiness concerning the independent, or indeed, the *a priori* existence of "empty space." The real aim of the elementary discussion given here is to show to what extent these doubts are overcome by the general theory of relativity.

The Concept of Space in the General Theory of Relativity

This theory arose primarily from the endeavour to understand the equality of inertial and gravitational mass. We start out from an inertial system S_1, whose space is, from the physical point of view, empty. In other words, there exists in the part of space contemplated neither matter (in the usual sense) nor a field (in the sense of the special theory of relativity). With reference to S_1 let there be a second system of reference S_2 in uniform acceleration. Then S_2 is thus not an inertial system. With respect to S_2 every test mass would move with an acceleration, which is independent of its physical and chemical nature. Relative to S_2, therefore, there exists a state which, at least to a first approximation, cannot be distinguished from a gravitational field. The following concept is thus compatible with the observable facts: S_2 is also equivalent to an "inertial system"; but with respect to S_2 a (homogeneous) gravitational field is present (about the origin of which one does not worry in this connection). Thus when the gravitational field is included in the framework of the consideration, the inertial system loses its objective significance, assuming that this "principle of equivalence" can be extended to any relative motion whatsoever of the systems of reference. If it is possible to base a consistent theory on these fundamental ideas, it will satisfy of itself the fact of the equality of inertial and gravitational mass, which is strongly confirmed empirically.

Considered four-dimensionally, a non-linear transformation of the four co-ordinates corresponds to the transition from S_1 to S_2. The question now arises: What kind of non-linear transformations are to be permitted, or, how is the Lorentz transformation to be generalised? In order to answer this question, the following consideration is decisive.

We ascribe to the inertial system of the earlier theory this property: Differences in co-ordinates are measured by stationary "rigid" measuring rods, and differences in time by clocks at rest. The first assumption is supplemented by another, namely, that for the relative laying out and fitting together of measuring rods at rest, the theorems on "lengths" in Euclidean geometry hold. From the results of the special theory of relativity it is then concluded, by elementary considerations, that this direct physical interpretation of the co-ordinates is lost for systems of reference (S_2) accelerated relatively to inertial systems (S_1). But if this is the case, the co-ordinates now express only the order or rank of the "contiguity" and hence also the dimensional grade of the space, but do not express any of its metrical properties. We are thus led to extend the transformations to arbitrary continuous transformations.[1] This implies the general principle of relativity: Natural laws must be covariant with respect to arbitrary continuous transformations of the co-ordinates. This requirement (combined with that of the greatest possible logical simplicity of the laws) limits the natural laws concerned incomparably more strongly than the special principle of relativity.

[1] This inexact mode of expression will perhaps suffice here.

This train of ideas is based essentially on the field as an independent concept. For the conditions prevailing with respect to S_2 are interpreted as a gravitational field, without the question of the existence of masses which produce this field being raised. By virtue of this train of ideas it can also be grasped why the laws of the pure gravitational field are more directly linked with the idea of general relativity than the laws for fields of a general kind (when, for instance, an electromagnetic field is present). We have, namely, good ground for the assumption that the "field-free" Minkowski-space represents a special case possible in natural law, in fact, the simplest conceivable special case. With respect to its metrical character, such a space is characterised by the fact that $dx_1^2 + dx_2^2 + dx_3^2$ is the square of the spatial separation, measured with a unit gauge, of two infinitesimally neighbouring points of a three-dimensional "space-like" cross section (Pythagorean theorem), whereas dx_4 is the temporal separation, measured with a suitable time gauge, of two events with common (x_1, x_2, x_3). All this simply means that an objective metrical significance is attached to the quantity

$$ds^2 = dx_1^2 + dx_2^2 + dx_3^2 - dx_4^2 \qquad (1)$$

as is readily shown with the aid of the Lorentz transformations. Mathematically, this fact corresponds to the condition that ds^2 is invariant with respect to Lorentz transformations.

If now, in the sense of the general principle of relativity, this space (cf. eq. (1)) is subjected to an arbitrary continuous transformation of the co-ordinates, then the objectively significant quantity ds is expressed in the new system of co-ordinates by the relation

$$ds^2 = g_{ik} dx_i dx_k \quad . \quad . \quad . \quad (1a)$$

which has to be summed up over the indices i and k for all combinations 11, 12, . . . up to 44. The term g_{ik} now are not constants, but functions of the co-ordinates, which are determined by the arbitrarily chosen transformation. Nevertheless, the terms g_{ik} are not arbitrary functions of the new co-ordinates, but just functions of such a kind that the form (1*a*) can be transformed back into the form (1) by a continuous transformation of the four co-ordinates. In order that this may be possible, the function g_{ik} must satisfy certain general covariant equations of condition, which were derived by B. Riemann more than half a century before the formulation of the general theory of relativity ("Riemann condition"). According to the principle of equivalence, (1*a*) describes in general co-variant form a gravitational field of a special kind, when the functions g_{ik} satisfy the Riemann condition.

It follows that the law for the pure gravitational field of a general kind must be satisfied when the Riemann condition is satisfied; but it must be weaker or less restricting than the Riemann condition. In this way the field law of pure gravitation is practically completely determined, a result which will not be justified in greater detail here.

We are now in a position to see how far the transition to the general theory of relativity modifies the concept of space. In accordance with classical mechanics and according to the special theory of relativity, space (space-time) has an existence independent of matter or field. In order to be able to describe at all that which fills up space and is dependent on the co-ordinates, space-time or the inertial system with its metrical properties must be thought of at once as existing, for otherwise the description of "that which fills up space" would have no meaning.[1] On the basis of the general theory of relativity, on the other hand, space as opposed to "what fills space," which is dependent on the co-ordinates, has no separate existence. Thus a pure gravitational field might have been described in terms of the g_{ik} (as functions of the co-ordinates), by solution of the gravitational equations. If we imagine the gravitational field, *i.e.* the functions g_{ik}, to be removed, there does not remain a space of the type (1), but absolutely *nothing*, and also no "topological space." For the functions g_{ik} describe not only the field, but at the same time also the topological and metrical structural properties of the manifold. A space of the type (1), judged from the standpoint of the general theory of relativity, is not a space without field, but a special case of the g_{ik} field, for which—for the co-ordinate system used, which in itself has no objective significance—the functions g_{ik} have values that do not depend on the co-ordinates. There is no such thing as an empty space, *i.e.* a space without field. Space-time does not claim existence on its own, but only as a structural quality of the field.

[1] If we consider that which fills space (e.g. the field) to be removed, there still remains the metric space in accordance with (1), which would also determine the inertial behaviour of a test body introduced into it.

Thus Descartes was not so far from the truth when he believed he must exclude the existence of an empty space. The notion indeed appears absurd, as long as physical reality is seen exclusively in ponderable bodies. It requires the idea of the field as the representative of reality, in combination with the general principle of relativity, to show the true kernel of Descartes' idea; there exists no space "empty of field."

Generalised Theory of Gravitation

The theory of the pure gravitational field on the basis of the general theory of relativity is therefore readily obtainable, because we may be confident that the "field-free" Minkowski space with its metric in conformity with (1) must satisfy the general laws of field. From this special case the law of gravitation follows by a generalisation which is practically free from arbitrariness. The further development of the theory is not so unequivocally determined by the general principle of relativity; it has been attempted in various directions during the last few decades. It is common to all these attempts, to conceive physical reality as a field, and moreover, one which is a generalisation of the gravitational field, and in which the field law is a generalisation of the law for the pure gravitational field. After long probing I believe that I have now found[1] the most

[1] The generalisation can be characterised in the following way. In accordance with its derivation from empty "Minkowski space," the pure gravitational field of the functions g_{ik} has the property of symmetry given by $g_{ik} = g_{ki}$ ($g_{12} = g_{21}$, etc.). The generalised field is of the same kind, but without this property of symmetry. The derivation of the field law is completely analogous to that of the special case of pure gravitation.

natural form for this generalisation, but I have not yet been able to find out whether this generalised law can stand up against the facts of experience.

The question of the particular field law is secondary in the preceding general considerations. At the present time, the main question is whether a field theory of the kind here contemplated can lead to the goal at all. By this is meant a theory which describes exhaustively physical reality, including four-dimensional space, by a field. The present-day generation of physicists is inclined to answer this question in the negative. In conformity with the present form of the quantum theory, it believes that the state of a system cannot be specified directly, but only in an indirect way by a statement of the statistics of the results of measurement attainable on the system. The conviction prevails that the experimentally assured duality of nature (corpuscular and wave structure) can be realised only by such a weakening of the concept of reality. I think that such a far-reaching theoretical renunciation is not for the present justified by our actual knowledge, and that one should not desist from pursuing to the end the path of the relativistic field theory.

A Reading Companion

This reading companion contains a series of commentaries on the basic ideas, concepts, and methods that are the building blocks of the theory of relativity—the special and the general. It should not be perceived as an alternative to Einstein's text. Moreover, the beauty and completeness of Einstein's booklet in presenting the two chapters of his relativity revolution does not depend on these commentaries. In fact, most editions of this book do not contain such auxiliary comments (to our knowledge, there is only one exception).[1] The aim of this reading companion is to help the reader by shedding additional light on the issues covered in the booklet, by tracing their evolution in the course of Einstein's thinking both before and after the booklet's publication, and by looking at them from a present-day perspective. We sometimes paraphrase Einstein's language but always connect it to something outside the text.

1 Einstein, Albert. *Relativity: The Special and the General Theory*, trans. Robert W. Lawson; introduction by Roger Penrose; commentary by Robert Geroch; with a historical essay by David C. Cassidy (New York: Pi Press, 2005).

PHYSICS AND GEOMETRY (§§ 1–2)

The first section addresses a central issue that appears throughout this book: the relation between physics and geometry. Almost everyone has encountered Euclid's geometry—with its authoritative appeal—at some point in his or her education. Indeed, for many centuries Euclidean geometry has served as the model for a true scientific theory. However, Einstein challenges the reader to question the confidence in the validity of geometrical assertions by examining the meaning of their "truth." During its long history, geometry has also repeatedly served as a reference point for articulating a philosophical understanding of scientific truth. Einstein developed his own views in close relation to his formulation of relativity theory.

He first critically addresses the view of geometry as a purely mathematical theory comprising propositions logically derived from a set of axioms, a view favored by contemporary mathematicians. The "truth" of such propositions then reduces to the "truth" of the axioms. However, what the truth of these axioms means remains unclear. This problem became particularly evident when one realizes that alternative axioms may be formulated that give rise to other, non-Euclidean geometries. Which are then the "true" axioms? This question cannot be decided on purely mathematical grounds.

Einstein explains the tendency to nevertheless ascribe truth to geometrical propositions by referring to their origin in experiences with real objects that underlie basic geometrical concepts, such as the concept of a rigid body. Rather than eliminating such concepts for the sake of logical closure of geometry, Einstein explicitly adds them to the conceptual edifice, thus turning geometry into a physical science where truth can be established by agreement with reality. He illustrates the usefulness of this addition by referring to examples from practical knowledge such as checking whether three places lie on a straight line by trying to align them along the line of vision. Similarly, Einstein refers to the habit of using a "practically rigid" body for geometrical measurements. He emphasizes the role of such practical experiences as a knowledge layer in its own right, rather than attempting to turn geometry into a physical science by means of an overarching foundational theory. In fact, the notion of a practically rigid body serves as a provisional mental model open to revision when further knowledge makes this necessary. Indeed, the concept of rigid body becomes problematic in the theory of relativity.

Einstein expressed these ideas in greater detail and with a decisive message about its significance in a lecture, "Geometry and Experience," delivered to the Prussian Academy of Science in January 1921.[2] It is worth summarizing its key argument here. Mathematics in general and geometry in particular have been embraced by humankind because of the need to gain information about the behavior of real objects. In fact, the word *geometry* means to measure the earth. Pure axiomatic geometry alone cannot make assertions about the behavior of real objects necessary for the process of surveying but must be supplemented with a relation to real objects, thus constituting a "practical geometry" in the sense described earlier.

Einstein concludes: "Geometry thus completed is evidently a natural science; we may in fact regard it as the most ancient branch of physics." He emphasizes the importance of this statement: "I attach special importance to

2 In CPAE vol. 7, Doc. 52, pp. 208–222, p. 211 is cited below.

the view of geometry which I have just set forth, because without it I should have been unable to formulate the theory of relativity." We shall substantiate this message in forthcoming commentaries.

To measure the distance between two points one needs a standard measuring rod and a method of defining the position of a point on a rigid body. To this end Einstein introduces, in section 2, the system of Cartesian coordinates, emphasizing that the system of coordinates has to be attached to a rigid body, and the position of a point is defined with respect to that body. Just as the intuitive notion of a rigid body was fundamental to Einstein's introduction of practical geometry, coordinates are introduced here as a generalization of the mental model of landmarks, such as the name of a place in a city that is familiar from everyday experience. Analogously, coordinates relate a place to a set of numbers obtained by measurements with the help of practically rigid bodies for whose displacements the laws of Euclidean geometry are assumed to hold. The basic notions in this section are quite simple and probably known to most readers. Still, Einstein takes great care in discussing these basic notions, because that simplicity will disappear and everything will have to be reexamined within the realm of general relativity.

MECHANICS AND SPACE (§§ 3–6)

Sections 3 through 6 elucidate the basic tenets of the description of motion in classical mechanics, culminating in the classical law of the addition of velocities. As in the previous two sections, Einstein discusses with great care the obvious and deceptively simple elements of this description, preparing the reader for the need to abandon some of these tenets in the transition to the special theory of relativity and even more so in general relativity.

He begins with the description of a motion and then proceeds to discuss the special case of inertial motion and the privileged reference frames of classical mechanics, which he designates the "Galileian system of co-ordinates." Next, he turns to the classical relativity principle to derive the law of the addition of velocities, which will serve as the starting point for demonstrating the need to go beyond the classical concepts of space and time in sections to come.

The description of motion of a material body in classical physics specifies the path of that body in space and the time elapsed during its motion at the different positions on this path. The preceding sentence contains the words *path*, *space*, *position*, and *time*, which represent concepts that require careful consideration. In particular, it is meaningless to ask about the path of the body in space without specifying a rigid body, called the *body of reference*, with respect to which this path is described. The mathematical description of this path requires a *system of coordinates* rigidly attached to the body of reference. The body of reference together with the system of coordinates is today commonly referred to as the *frame of reference*.

One of the most remarkable achievements of Newtonian mechanics is the insight that there is a special class of particularly simple, "undisturbed" objects in nature that move along a special class of particularly simple paths. The simple objects may be characterized as *free particles*, and the simple motions are what are now called *inertial motions*, in this case, uniform motions along straight lines. Deviations from these simple motions are attributed to the intervention of a force. Newton's formulation of the principle of inertia refers, just as Einstein's formulation does, to all bodies: "Every body perseveres in its state of being at rest or of moving uniformly straight forward, except insofar as it is compelled to change its state by forces impressed." Bodies subject to forces thus no longer qualify as simple objects following simple inertial paths. But using the concept of force in the definition of inertial motions appears to be circular reasoning when forces are defined only as deviations from such motions. To avoid this circularity Einstein characterizes a body in inertial motion as a body sufficiently removed from other bodies. However, this characterization is also problematic, since it conflicts with the fact that gravitation, in particular, has an infinitely long range and cannot be shielded. A profound solution to this problem, ultimately due to the inseparability of gravitational and inertial aspects, will be offered by general relativity.

In the preceding section, Einstein emphasized that the path of an object depends on the system of coordinates with respect to which it is described. An object moving with constant velocity along a straight line with respect to the ground will follow a very different path in a system of coordinates attached to a merry-go-round. A system of coordinates in which the principle of inertia can be expressed in terms of uniform motion along straight lines is called a *Galilean system of coordinates*, commonly referred to as an *inertial frame of reference*. Such reference frames offer a particularly simple formulation of the laws of classical mechanics. But how do we choose such a frame of reference? Historically, this was a much-discussed question. Einstein argues that we must first pick a specific case of motion that we can reasonably characterize as inertial motion, such as the motion of fixed stars, and then look for a coordinate system in which this motion can be described as a uniform motion along a straight line. This precludes using a system rigidly connected to the earth, from which the fixed stars appear to perform daily rotations.

The Galilean coordinate system does not single out one preferred inertial reference frame that may be considered "at rest." Rather, there is an entire class of them. Each reference frame that moves uniformly along a straight line with respect to a given inertial frame also qualifies as an inertial frame with respect to which the principle of inertia holds in the simple form described earlier. More generally, the same general laws of nature hold in each of these inertial frames, a statement that Einstein refers to as the principle of relativity in the restricted sense. The laws of nature are left unchanged in going from one inertial frame to another, just as the translation of a figure in Euclidean space leaves its geometrical properties unchanged, thus exhibiting a symmetry of this space. Similarly, the principle of relativity constitutes a symmetry of Newtonian space, because the laws of mechanics do not depend on the location and direction in space. This is a very important basic principle, so a careful analysis of it will be reproduced at the beginning of the transition to general relativity. The "Restricted Sense" in the section title refers to the fact that the discussion concerns only Galilean systems of coordinates, namely, inertial frames of reference.

The validity of the relativity principle was well established for the laws of Newtonian mechanics. The considerable extension of the range of experience of physics during the nineteenth century raised the question whether this principle also held for the domains of optics and electromagnetism, in particular since it turned out to be impossible simply to reduce these domains to the laws of mechanics. It was therefore quite conceivable that even such a fundamental principle of mechanics as the relativity principle would not hold for them. This conjecture also appeared plausible because the development of optics and electrodynamics crucially hinged on the concept of an *ether*, a hypothetical invisible medium that permeates the entire physical space and serves as a carrier of light and other electromagnetic fields. A stationary ether would naturally provide a unique frame of reference at absolute rest.

Here Einstein does not mention the introduction and the general acceptance of the concept of an ether in prerelativistic physics. Rather, he limits himself to giving two general reasons why it is nevertheless to be expected that the principle of relativity is not restricted to the domain of mechanics but also applies to these other domains. His first argument is the great success of mechanics in spite of its limitations and the generality of the principle of relativity that makes it improbable that it holds only for mechanics. The second reason refers to the effect of the motion of the earth around the sun. If the relativity principle should not hold for all of physics and if instead there were a privileged frame at rest, then space would be anisotropic with respect to motions in different directions. But in spite of many efforts, such anisotropy has never been observed. Though not mentioning them explicitly, Einstein refers here to experiments, such as the famous Michelson-Morley interferometry experiment, that have failed to yield any observable effects of the motion of the earth on the propagation of light.

Einstein's recapitulation of the essence of classical Newtonian mechanics has a simple, seemingly trivial consequence, which he presents in section 6, the addition theorem for velocities. If a motion of a physical object is considered first with respect to one reference frame and then with respect to another reference frame that is itself in motion with respect to the first frame, then the motion of the physical object with respect to the second reference frame is obtained by a combination of the two motions. If the motions take place along the same line and in the same direction, the resulting velocity is, according to classical physics, simply the sum of the velocities of the two motions. This seemingly obvious conclusion is, however, in conflict with a fundamental principle of optics and electrodynamics, a conflict to which Einstein dedicates the subsequent section.

LIGHT PROPAGATION AND TIME (§§ 7–9)

In the next three sections Einstein takes the decisive step from classical to relativistic physics. He first explains that this step has become necessary because of a fundamental conflict within classical physics and then shows how this conflict can be resolved by giving up the classical concept of time. In the subsequent sections he will similarly reexamine the classical concept of space.

The starting point is an apparently simple physical fact, the constancy of the speed of light. This constancy holds for all colors of light. Astronomical observations have furthermore confirmed that the speed of light is independent of its source.[3] The propagation of light is a typical borderline problem between two domains of physics—optics and

3 De Sitter, Willem. Ein astronomischer Beweis für die Konstanz der Lichgeschwindigkeit. *Physik. Zeitschr.*, 14, 429 (1913); Über die Genauigkeit, innerhalb welcher die Unabhängigkeit der Lichtgeschwindigkeit von der Bewegung der Quelle behauptet werden kann. *Physik. Zeitschr.*, 14, 1267 (1913); A proof of the constancy of the velocity of light. *Proceedings of the Royal Netherlands Academy of Arts and Sciences* 15 (2): 1297–1298 (1913); On the constancy of the velocity of light. *Proceedings of the Royal Netherlands Academy of Arts and Sciences* 16 (1): 395–396 (1913).

mechanics—that may lead to conceptual conflicts. How is the constancy of the speed of light to be understood from the point of view of the laws of motion previously discussed?

Suppose that a light beam is sent along the carriage of a moving train, in the direction of its motion, with velocity c. The speed of light with respect to the carriage moving in the same direction is accordingly diminished by the speed of the train. With respect to an observer in the train the beam will thus propagate with velocity c minus the velocity of the train. This result follows from the consideration of the addition of velocities in the preceding section, which is a direct consequence of the classical notions of space and time, and in particular of the assumption that there is an absolute time running in the same way in all reference frames. But this conclusion violates the principle of relativity, according to which the fundamental laws of nature—and hence also the law of the constancy of the velocity of light—should be the same in all inertial frames.

Einstein now explains the options for resolving this dilemma. In view of the fundamental character of the relativity principle explained in section 5, it seems plausible to maintain this principle and look for a modification of the law of light propagation that makes it compatible with the relativity principle. Before Einstein published his paper "On the Electrodynamics of Moving Bodies"[4] in 1905—establishing what was later to be called *special relativity*—he had himself looked in a similar direction, trying to create a new theory of light compatible with the relativity principle. But as he explains next, the success of the established theory of electrodynamics formulated by

4 CPAE vol. 2, Doc. 23, pp. 140–171.

the Dutch physicist Hendrik Antoon Lorentz (1853–1928) was so overwhelming that it turned out to be impossible to find a convincing alternative to it. However, this theory is based on the premise that there is a privileged frame of reference in which the speed of light in a vacuum is a constant.

The extraordinary success of Lorentz's theory in explaining all optical and electromagnetic phenomena had thus underscored the dilemma, turning the constancy of the speed of light into an undeniable fundamental law of nature that was seemingly in conflict with the relativity principle. This was the situation in which the young Einstein realized that the solution of this profound dilemma was to be found not in an alternative theory of electromagnetism and optics but in new concepts of space and time. Our text evidently recapitulates Einstein's own investigative pathway.

It was his reading of philosophical texts, in particular the works of the philosophers David Hume (1711–1776) and Ernst Mach (1838–1916), and discussions with friends—in particular with the engineer Michele Besso—that helped Einstein reach this other level of reflection. The lesson that he derived is that our concepts of space and time are human constructions that should be examined in relation to experience and that such an examination may lead to a revision of these concepts when new experiences such as those embodied in the laws of electrodynamics are taken into account.

Einstein's 1905 paper,[5] offering the first formulation of the special theory of relativity, begins with a detailed analysis of the meaning of the statement "the train arrives here at 7 o'clock." This analysis requires a careful examination of the meaning of time and its measurement, and leads

5 CPAE vol. 2, Doc. 23, pp. 140–171.

to the conclusion that the basic notion of simultaneity is relative to the reference frame chosen. If an observer in one inertial frame of reference decides—by a well-defined method described in sections 8 and 9—that two events occur simultaneously, then these two events will not be simultaneous in another frame of reference moving with constant velocity with respect to the first one. A new interpretation of the concept of simultaneity is required because of the finiteness of the speed of light through which observers receive information about events at distant locations.

> Referring to the meaning of simultaneity much later in his *Autobiographical Notes*, Einstein commented on how difficult it was to get rid of the belief in the absolute nature of simultaneity: "To recognize clearly this axiom and its arbitrary character already implies the essentials of the solution of the problem. The type of critical reasoning required for the discovery of this central point [the relativity of simultaneity] was decisively furthered, in my case especially, by the reading of David Hume's and Ernst Mach's philosophical writings."[6]

The relative nature of simultaneity undermines the basis of the addition law of velocities that leads to the conflict between the constancy of the speed of light and the relativity principle. It now becomes conceivable that an alternative addition law can avoid this conflict and lead instead to the conclusion that the speed of light as measured in the train is the same as that measured by the observer on the embankment.

6 *Autobiographical Notes*, p. 51.

LIGHT PROPAGATION AND SPACE (§§ 10–12 AND APPENDIX 1)

In the following three sections Einstein explains the basic properties of space and time underlying his theory of special relativity to resolve the conflict between the relativity principle and the postulate of the constancy of the speed of light. These basic properties are described with the help of the Lorentz transformations, which are derived in appendix 1. In the main text Einstein focuses on their application to the behavior of moving rods and clocks.

The measurement of the distance between two specific points on a train by a passenger on the train is a simple matter. The passenger has only to use a measuring rod of unit length and count how many times this rod fits into the interval between these two points. The situation is more complicated if the measurement is to be performed by an observer on the embankment, with respect to which the train is moving at constant velocity v. First, one has to define the procedure by which such a measurement is performed. Einstein specifies this procedure as follows. An observer on the embankment has to mark the points A and B on the embankment that coincide with the two specific points on the passing train. The critical element is that these two coincidences must take place at the same instant of time t measured by a clock on the embankment. Then, after the train has passed, the observer may measure, in the ordinary way, the spatial distance between them.

In classical mechanics the two measurements of the length of a moving rod—by an observer moving with the rod and by a stationary observer—yield the same result. Likewise, the time interval between two events is independent of the state of motion of the frame of reference.

These two basic tenets of classical mechanics lead to the contradiction between the principle of relativity and the constancy of the velocity of light, discussed in section 7. How should the two tenets be modified to remove the incompatibility between these two principles? The answer is given by the Lorentz transformation, which defines the mathematical relation between the spatial and time coordinates x, y, z, t in one reference frame K and the coordinates x', y', z', t' of the same event with respect to another reference frame K' in uniform motion with respect to the former. The mathematical form of the Lorentz transformation is presented in section 11, in comparison with the Galilean transformation of classical mechanics.

The following are the main consequences of the Lorentz transformation:

The constancy of the velocity of light– If a light signal is sent, in reference frame K, from the origin of the coordinate system ($x = 0, y = 0, z = 0$) at time $t_0 = 0$ along the x axis, then at time t it reaches the point $x = ct$. In a different reference frame K', the coordinates x, t are transformed into x', t', so that $x' = ct'$. Thus, the light signal propagates in reference frame K' with the same velocity c.

The length contraction– If the length of a rod in its own reference frame is L, then the length of that rod measured in a reference frame moving with velocity v with respect to the rod is obtained by multiplying L with $\sqrt{1-v^2/c^2}$. Thus, the length of a rod measured by a moving observer is shortened with respect to the rod at rest.

The time dilation– Let the time interval between two events taking place at the same location in one reference system be equal to t as measured by a clock within this reference system. The time interval between these events measured by a clock moving with velocity v with respect to the first reference frame is $\frac{t}{\sqrt{1 - v^2 / c^2}}$. Thus, the rate of a moving clock is slower than that of a clock at rest.

The Lorentz transformation was introduced, before Einstein's special theory of relativity, by the Dutch physicist Hendrik Antoon Lorentz in the context of his studies of electromagnetism. In 1905, Einstein recognized that this transformation is implied by the requirement that the relativity principle also apply to electromagnetism. Maxwell's equations, which describe all electromagnetic phenomena, preserve their form (are covariant) in different inertial frames of reference under the Lorentz transformation.

The shortening of the length of a moving rod is known as the *Lorentz-FitzGerald contraction*. It was hypothetically introduced by Hendrik Lorentz and the Irish physicist George FitzGerald (1851–1901) to explain the failure to detect the motion of the earth through the ether and thereby to rescue the hypothesis of the ether as a stationary reference frame, which at that time was one of the generally accepted cornerstones of physics. We shall return to the last two points in the commentary on section 16.

PHYSICS IN RELATIVISTIC SPACE AND TIME (§§ 13–16)

In the following four sections Einstein reviews the evidence in favor of his theory from well-known classical experiments, such as the Fizeau experiment, to its revolutionary implications such as the equivalence of mass and energy.

Einstein was led to the discovery of the special theory of relativity by his attempt to reconcile the principle of relativity with the principle of the constancy of the velocity of light. The incompatibility of these two principles stems from the fact that the principle of relativity in classical physics implies that if a person is moving along a train with velocity w with respect to the train, and the train is moving with velocity v with respect to the embankment, then an observer on the embankment will measure the velocity of the person on the train as $W = v + w$. If we replace the person on the train by a beam of light propagating with velocity c, then this velocity, seen by an observer on the embankment will be $W = v + c$. This result violates the principle that the velocity of light should remain constant in all reference frames. Thus, to resolve this incompatibility a new prescription for the addition of velocities is required. Such a prescription is implied by the Lorentz transformation, and its mathematical form is given by formula B on page 51. A simple manipulation of this new addition law for velocities implies that the principle of the constancy of the velocity of light is valid.

Einstein now turns to the Fizeau experiment, which he perceives as crucial evidence for the new addition law following from his theory. In the middle of the nineteenth century, the physicist Hippolyte Fizeau (1819–1896) performed an experiment to study the effect of movement of a medium on the velocity of light. In this experiment a light signal is sent through a moving liquid, which immediately raises the question, how is the velocity of light affected by the moving medium? and thus concerns the law of addition of velocities. In a fluid medium light propagates at a lower velocity determined by the so-called refractive index of the fluid n. The parameter n, which can be determined from the optical properties of the fluid, is equal to the ratio of the velocity of light in vacuum and in the fluid. Fizeau measured the speed of light propagating in flowing water. If the hypothetical ether were totally dragged along, the result would be the sum of the flow velocity of the water plus the velocity of light in water, determined by the refractive index of water. However, the observed value was significantly lower than this. At the time, this result was interpreted with the help of the assumption that the ether carrying the light signal was only partially dragged along by the moving liquid. Einstein quotes this result as conclusive confirmation of the relativistic velocity addition law.

The heuristic value of the special theory of relativity lies in its specification of a mathematical criterion that all systems of equations describing laws of nature must satisfy. The mathematical expressions describing such laws depend on space and time coordinates. The coordinates change under a Lorentz transformation from one reference frame to another, but the mathematical formulation of a physical law must remain unchanged; the laws must be "covariant" under the Lorentz transformation, because the principle of relativity states that such laws are the same in all inertial frames of reference. Maxwell's equations meet this criterion, which is not surprising, because the Lorentz transformation was constructed for this purpose. However, an inspection of Newton's law of motion reveals that it does not and therefore has to be modified to conform to this criterion. The appropriate modification renders new results.

The Equivalence of Mass and Energy

The most important and best-known result of relativistic mechanics is that the *inertial mass* of a body, which determines the acceleration of a body caused by a specific force, varies with a change in the energy of the body and can be regarded as a measure of its energy content. Einstein had initially overlooked this consequence of the special theory of relativity in his seminal paper "On the Electrodynamics of Moving Bodies"[7] and published it only a few months later in the short follow-up paper "Does the Inertia of a Body Depend on Its Energy Content?"[8] He repeats the argument in the present booklet, showing that a body receiving energy E in the form of radiation increases in mass by E/c^2.

> In the summer of 1905, Einstein wrote to his friend Conrad Habicht, with whom he often shared his ideas in those days: "A consequence of the study on electrodynamics did cross my mind. Namely, the principle of relativity, in association with Maxwell's fundamental equations, requires that mass be a direct measure of the energy contained in a body; light carries mass with it. A noticeable reduction of mass would have to take place in the case of radium. The consideration is amusing and seductive; but for all I know, God Almighty might be laughing at the whole matter and might have been leading me around by the nose."[9]

One immediate consequence of this result, which Einstein emphasized on several occasions, including here, is that in classical physics there are two independent conservation laws—conservation of mass and conservation of energy. Since mass is a form of energy, changes in mass are part of the energy balance in any physical process. There-

7 CPAE vol. 2, Doc. 23, pp. 140–171.

8 CPAE vol. 2, Doc. 24, pp. 172–174.

9 Einstein to Conrad Habicht, May 1905, CPAE vol. 5, Doc. 28, p. 21.

fore, in relativistic mechanics there is only one conservation law.

The equivalence between mass and energy is represented by the mathematical expression $E = mc^2$, probably the most famous combination of five symbols in the history of mankind. It appears on commercial products and on postal stamps of almost every country on Earth. Since the velocity of light is such a large number, and its square is still much larger, the expression indicates that a small amount of mass is equivalent to a large amount of energy. This energy corresponds to that of a particle of mass m at rest. For a particle moving with velocity v, the energy is given by the second expression on page 295, which for small v reduces to the left-hand side of equation [21], which is the sum of the energy content of the mass m and the classical kinetic energy of a moving particle of mass m.

The validity of the equation $E = mc^2$ could not be directly demonstrated at that time, but was confirmed for the first time in a nuclear reaction in 1932, in the Cavendish Laboratory in England, by the British and Irish physicists John Cockroft and Ernest Walton, respectively. They developed the first particle accelerator and used it to smash accelerated protons into a lithium nucleus target. The collision produced two helium nuclei (so-called alpha particles). The sum of the masses of a proton and a lithium nucleus exceeds the sum of the masses of two alpha particles. The mass deficit was converted into a huge kinetic energy of the alpha particles.

Lord Ernest Rutherford, one of the founding fathers of nuclear physics and head of the Cavendish Laboratory, described this experiment in 1933: "We might in these processes obtain very much more energy than the proton supplied, but on the average we could not expect to obtain energy in this way. It was a very poor and inefficient way of producing energy, and anyone who looked for a source of power in the transformation of the atoms was talking moonshine. But the subject was scientifically interesting because it gave insight into the atoms."[10] Six years later this "moonshine talk" led to the discovery of nuclear fission and only 12 years later to the destruction of two cities in Japan by an atomic bomb.

One nuclear process in which mass is converted into energy affects humankind most. In the core of the sun, four protons in a chain of nuclear reactions are transformed into one alpha particle. The mass of four protons is larger than the mass of an alpha particle. The mass lost in this process is the source of solar energy and, therefore, directly and indirectly the source of energy and life on Earth.

Is the Special Theory of Relativity Supported by Experiment?

Einstein first emphasizes that anything that is consistent with the Maxwell-Lorentz theory of electromagnetism also supports special relativity, because the latter emerged from that theory. As an example he mentions the effects of Earth's motion on the light of fixed stars. Although these effects can, in principle, be explained by classical theory, the explanation in terms of special relativity is much simpler. In addition, special relativity also offered new and straightforward explanations for effects that the older theory could account for only by introducing artificial hy-

10 "The British Association: Breaking Down the Atom." *Times* (London), September 12, 1933.

potheses, such as the contraction of a body due its motion through the ether. Einstein illustrates this with two classes of experimental findings, one referring to particle beams, the other to interference experiments with light.

Extensive experiments were performed on beams of electrons provided by the so-called beta radiation, emitted by certain radioactive elements, and by cathode rays, which are electron beams emitted by a negatively charged plate (the cathode) that propagate in a vacuum tube to a positively charged anode. By the time of the publication of this booklet there was no discrepancy between these experiments and the predictions of the special theory of relativity. These predictions essentially coincided with those of Lorentz, who needed, however, to introduce the contraction of the electron in its direction of motion. This had not been the case 10 years earlier. Most experiments had been inconclusive, and one experienced and respected experimental physicist, Walter Kaufmann, had published results that contradicted Einstein's predictions and even suggested alternative equations of motion of the electrons. Kaufmann's result generated intensive debates. Einstein doubted their validity and particularly objected to the proposed equations of motion. Although he accepted experiment as the ultimate test of a theory, in this case Einstein had such confidence in the theoretical basis of special relativity that he believed the experiments would ultimately fall in line, as they did.

At the end of the nineteenth century and the beginning of the twentieth a number of empirical results seriously challenged the accepted framework of classical physics, and prominent physicists attempted to resolve these difficulties. With the exception of Einstein, most physicists did not want to give up the basic laws and hypotheses of classical physics but were ready to adopt additional hypotheses that did not follow from known laws of physics and were difficult to justify. The best known of those empirical studies, published in 1887, was the experiment performed by the U.S. physicists Albert Michelson (1852–1931) and Edward Morley (1838–1923) to detect the motion of Earth with respect to the ether. The hypothesis of the existence of such an invisible medium had been introduced into classical physics for two purposes. It was supposed to serve as the medium in which the electromagnetic waves, predicted by Maxwell's equations, could propagate, just as sound waves propagate through air and ocean waves advance on the surface of water. The ether also served as an absolute reference frame with respect to which Maxwell's equations were valid. Under the Galilean transformation of classical physics they assume a different form in every reference frame moved with respect to that frame.

If there were a motion between Earth and an ether, there should be an "ether-wind" that would affect the velocity of light depending on the direction of its propagation, just as a swimmer moves faster or slower depending on whether he or she is moving with or against the current. The Michelson-Morley experiment was designed to measure such an effect. A beam of light was split into two beams moving in different directions, and after covering an equal distance they finally hit the same target. It was expected that if the two beams traveled the same distance along different paths through an ether, one parallel to the direction of motion of the Earth and the other perpendicular to it, they should move at different velocities and would therefore hit the final target at slightly different times. The remarkable result was that no such time difference could be detected. The accuracy of the measurement was such that it left no doubt about the validity of this conclusion.

To rescue the premises of classical physics, Lorentz and FitzGerald assumed that the length of the experimental apparatus moving with the Earth with respect to the ether is shortened by an amount that compensates for this time difference, as was mentioned in the commentary on section 12. Lorentz attributed this contraction of length to some hypotheses about the properties of charged matter (electrons). In Einstein's theory of special relativity the result follows naturally from the coexistence of the principle of relativity with the principle of the constancy of light velocity. In this theory there is no need for the ether concept.

> Einstein did not mention the Michelson-Morley experiment explicitly in his first paper on special relativity, but he alluded to it saying that "the failure of attempts to detect a motion of the earth relative to the "light medium" was among the causes which led him to the conclusion "that not only in mechanics, but also in electrodynamics as well, the phenomena do not have any properties corresponding to the concept of absolute rest."[11]

THE WORLD OF FOUR DIMENSIONS (§ 17 AND APPENDIX 2)

Hermann Minkowski (1864–1909) was a professor of mathematics at the Federal Institute of Technology in Zurich when Einstein was a student there and attended several of his courses. In 1908, he showed that Einstein's special

11 CPAE vol. 2, Doc. 23, p. 140.

theory of relativity could be understood geometrically as a theory of four-dimensional space-time.

We have already used the notion of an *event* to describe a physical occurrence characterized by its location and the time of its occurrence. The location in space is defined by three numbers (coordinates): x, y, z. Thus, space forms a three-dimensional continuum of all its points. The specification of an event requires an additional parameter (coordinate), its time t. In classical physics, time is absolute and is independent of the space coordinates. Therefore, in classical physics there is no advantage to treating events described by the four parameters x, y, z, t as points in a four-dimensional space-time. Instead, there is a three-dimensional space continuum and an independent one-dimensional time continuum.

The situation is different in special relativity. Suppose that the coordinates of an event in one coordinate system are x, y, z, t. The time t' of this event observed from another inertial system depends on both its time and space coordinates in the initial system. The essence of the Lorentz transformation is that this mixing of spatial and time coordinates makes it convenient to combine them into a single four-dimensional space-time. In Minkowski's formulation the time coordinate is always multiplied by the velocity of light c, so that ct has the dimension of distance, like the spatial coordinates. Moreover, it is multiplied by the imaginary number $i = \sqrt{-1}$. The four-dimensional continuum x, y, z, ict, to which Minkowski referred as the *world* and to which we refer as *space-time*, is essentially a four-dimensional geometrical space, resembling the ordinary Euclidean space. The behavior of a particle is described by a trajectory of points in space-time called the *world-line* of the particle. For example, a particle at rest is described by a world-line parallel to the time axis, and the collision of two particles is represented by the intersection of two world-lines.

Minkowski's four-dimensional space-time is equipped with a "metric" instruction that is employed to measure the "distance" between two events. The square of this distance is simply the square of the time separation between the two events (multiplied by c^2) minus the square of their spatial separation. This formula may be compared with the familiar metric instruction to measure the distance between two points in three-dimensional space: sum the squares of the separations between the spatial coordinates x, y, z, which corresponds to the extension of Pythagoras's theorem to three dimensions. Observers moving at constant velocity with respect to each other may compute this value using rods and clocks to perform position and time measurements in their frame of reference, and they will get the same result. In other words, the "distance" between two events is invariant under Lorentz transformations between different Galilean coordinate systems.

It took Einstein some time to appreciate Minkowski's geometrical formulation of the theory of special relativity as an interesting and useful contribution. Einstein became convinced of its basic importance only around 1912, during his search for a relativistic theory of gravitation. Minkowski's formulation of the theory became the framework of its later development and led Einstein to his theory of general relativity. In the first paragraph of Einstein's seminal article "The Foundation of the General Theory of Relativity," published in March 1916, he writes: "The generalization of the theory of relativity has been facilitated considerably by Minkowski, a mathematician who was the first one to recognize the formal equivalence of space coordinates and the time coordinate, and utilized this in the construction of the theory."[12]

FROM SPECIAL TO GENERAL RELATIVITY

The transition from the special to the general theory of relativity merits a brief discussion of the relationship and comparison between these two theories. If Einstein had not formulated the special theory in 1905, it would have probably been discovered, sooner or later, in one way or another. Results of experiments called for a new understanding of the notions of electrodynamics, the basic ideas were in the air, and the Lorentz transformation, though interpreted differently, was known. The same cannot be said about the general theory. There was little on the agenda of physics or of astronomical observations that required the bold extension of the special theory of relativity undertaken by Einstein. For him it was an intellectual necessity, and it could not have been accomplished, at that time, without his unique and original way of thinking.

The generalization of special relativity has been described as the quest for incorporating accelerated frames of reference into the theory. Such a description is misleading. Newtonian mechanics and special relativity can also be observed from an accelerated frame of reference, but the laws of physics are then more complicated. Einstein struggled to find a theory in which the physical laws would be the same in accelerated frames of reference. What he actually discovered was a theory in which gravitation and inertia are recognized as aspects of the same field. According to the equivalence principle, gravity is of the same nature as iner-

12 CPAE vol. 6, Doc. 30, p. 146.

tial forces, and the study of gravity naturally leads to non-inertial frames of reference. Thus, what is usually called the general theory of relativity is essentially a relativistic theory of gravity and inertia.

Historically, the special and general theories of relativity were developed and presented as two separate theories. At present they are perceived as two elements of one theory of relativity distinguished by the presence of the gravitational field. This perception is even more natural in Minkowski's space-time description. In this representation, special and general relativity are distinguished by the geometrical structure of the underlying space-time, which is characterized by its curvature everywhere. In this single theory, events are described in a possibly curved space-time, and the theory reduces to special relativity in the limit of flat space-time geometry.

This point will become more evident in the forthcoming commentaries.

GRAVITATION AND INERTIA (§§ 18–21)

In the present account of the transition from the special to the general theory, Einstein returns to a careful analysis of the *special* principle of relativity introduced and discussed in section 5, to remove the restriction implied by the word *special* and to extend it to systems of coordinates attached to rigid bodies in an arbitrary state of motion. A natural formulation of a *general* principle of relativity would read: all bodies of reference are equivalent for the description of the laws of nature, whatever may be their state of motion. Such a generalization seemed to Einstein to be an intellectual necessity. He writes: "Since the introduction of the special principle of relativity has been justified, every intellect which strives after generalization must feel the temptation to venture the step towards the general principle of relativity." While the very few "intellects" who were actually tempted to venture that step worked at the margin of physics and remained unsuccessful, Einstein alone persistently followed his intuition from 1907 until he formulated his general theory of relativity in 1915.

A person riding on a train can sense the difficulty of such a generalization. If a sudden acceleration is applied to the train, the person will be subject to a tilt backward or forward. Thus the mechanical behavior of bodies under such circumstances is different from the behavior in a Galilean system of coordinates. It is then natural to attribute an absolute physical reality to accelerated motion.

Einstein's attempt to cope with this apparent difference between uniform and accelerated motion led him to his theory of general relativity, which at the same time is a relativistic theory of the gravitational field. The concept of a *field* was developed by James Maxwell in the context of electromagnetism. In his *Autobiographical Notes*, Einstein recalls: "The most fascinating subject at the time that I was a student was Maxwell's theory. What made this theory appear revolutionary was the transition from action at a distance to fields as the fundamental variables."[13] The gravitational field has a peculiar property. Unlike in the case of an electric or magnetic field, in a gravitational field bodies of any size or material composition, starting from rest or uniform motion, will move at the same acceleration. This is one of the basic principles of classical physics, established by Galileo in his lifelong studies of falling bodies. This principle implies that the *inertial mass* of a body is always

13 *Autobiographical Notes*, p. 31.

equal to its *gravitational mass*, although conceptually the two masses are distinct. The inertial mass determines the acceleration of a body caused by a given force, while the gravitational mass determines the force exerted on a body by a given gravitational field. The equivalence of these two properties of a massive body was known in mechanics, and its validity had already been demonstrated empirically with great accuracy in Einstein's time, but its significance had not been explored. Only Einstein interpreted it as a basic principle and adopted it as a cornerstone of his general theory of relativity.

> In his brief introduction to the booklet Einstein states that it is intended for readers "who are not conversant with the mathematical expressions of theoretical physics." (p. 10) Nevertheless, he exposes the readers to the mathematical formulation of the Lorentz transformation and some of its consequences. In contrast, he refrains from mathematical formalism when, in section 19, he demonstrates that the gravitational field imparts the same acceleration to all objects. Simpler mathematical expressions would be needed in this case, but he uses relations between physical concepts expressed in words rather than in mathematical symbols.

Einstein took the equality of the inertial and gravitational mass of a body more seriously than others and formulated his famous *equivalence principle*. An observer in a closed box in outer space, far away from all heavenly bodies, will not feel a gravitational field in his or her immediate surroundings. Suppose that the box is made to move "upward" with constant acceleration. The person in the box, as demonstrated in section 20, has no way to decide if the effects he or she observes in the box are caused by a uniform acceleration of the box or by a gravitational field exerting a gravitational force in the opposite direction. Likewise, the passenger on a train who feels a tilt backward when the train is suddenly accelerated may assume that the train is at rest but a gravitational field has suddenly been applied to the system.

In 1922, Einstein delivered a lecture at the University of Kyoto titled "How I Created the Theory of Relativity." There he recalled that "I was sitting in a chair in the Patent Office in Bern when all of a sudden I was struck by a thought: 'If a person falls freely, he will certainly not feel his own weight.' I was startled. This simple thought made a really deep impression on me. My excitement motivated me to develop a new theory of gravitation."[14] Later in life, Einstein referred to this revelation as the "happiest thought" of his life.[15] This idea led him to conclude that gravitation can in special cases be eliminated by a transition to an accelerated frame of reference and to formulate his "equivalence principle."

After having established the equivalence between gravitation and inertia, Einstein returned to his basic criticism of classical mechanics and special relativity. He, like Mach, had always been disturbed by the special status of reference bodies with respect to which basic laws of physics hold and other reference frames with respect to which they do not. Einstein found it particularly disturbing that two physical situations can be distinguished by a single effect for which

14 Kyoto Lecture. In CPAE vol. 13 (German edition), Doc. 399, p. 638.

15 CPAE vol. 7, Doc. 31, p. 136.

no explanation is given. He illustrates this dilemma with the example of two identical pans with steam emerging from only one of them. In this context Newton considered the example of a rotating bucket filled with water in which effects occur that are not observed in a bucket at rest. He was also concerned by this distinction, but he did not explain it as an effect of specific physical causes but rather as evidence for absolute space. Einstein, however, states: "no person whose mode of thought is logical can rest with this condition of things." (p. 85) In fact, there were many people with a logical mode of thought who could. Einstein could not.

ACCELERATION, CLOCKS, AND RODS (§§ 22–23)

In sections 22 and 23 Einstein derives far-reaching consequences for the extension of his equivalence principle to a general principle of relativity, that is, from his insight that gravitation can, to some extent, be simulated by the effects occurring in an accelerated frame of reference. He was aware, of course, that not all gravitational fields can be generated in this way. Nevertheless, this heuristic strategy allowed Einstein to anticipate fundamental properties of a relativistic theory of gravitation even before explicitly formulating this theory, by cleverly combining previously acquired knowledge of classical mechanics and special relativity. Indeed, these conclusions are based on a combination of the equivalence principle with the results of special relativity. Einstein's exposition is not just a didactic device to make the subject more understandable to lay readers but actually corresponds to his own path of discovery. In 1907 Einstein formulated the equivalence principle and immediately inferred from it, for instance, that light must curve in a gravitational field. Later, in 1912, he realized that, in general, Euclidean geometry no longer holds when a gravitational field is present.

Einstein begins his considerations by comparing two reference frames, an inertial, "Galileian" system of reference and a frame accelerated with regard to it. According to the principle of inertia, a body not subject to any forces will move with uniform motion along a straight line if considered from the inertial system. In general, this straight line will become a curvilinear path if considered from an accelerated frame of reference. This insight may appear to be trivial, but when reformulated with the help of the appropriate mathematical formalism, it amounts to the discovery that such force-free motion within a gravitational field can be described by a so-called geodetic line. The concept of a straight line is thus generalized to a curved geometry, representing the straightest possible path within such geometry. Einstein realized this only in 1912 when he became aware of the connection between the problem of gravitation and geometry. He does not mention this mathematical insight in the present popular exposition.

Instead, he draws another conclusion from the general principle of relativity with immediate physical consequences: if light rays are curved when considered from an accelerated frame of reference, they must also be curved in a gravitational field. Einstein points out that this gravitational light deflection might be observed during a solar eclipse, a prediction that would be confirmed two years after his booklet first appeared. He also takes the opportunity to address the criticism of his opponents who maintain that general relativity contradicts special relativity—for which the constancy of the speed of light is a fundamental principle—while a curvature of light rays seems to indicate that the speed of light cannot be constant. Einstein addresses the criticism generically, using the example of the relation between electrodynamics and electrostatics to explain how one theory generalizes another theory by incorporating it as a limiting case. Overcoming the critics' objection requires careful consideration of how clocks and rods behave in a gravitational field and is dealt with in the subsequent section (23).

Einstein concludes section 22 by indicating that the gravitational field equation itself may be obtained from a generalized principle of relativity. Although this principle takes the special gravitational fields that can be generated by accelerated motion as its starting point, Einstein nevertheless claims that the general law can be derived from them. The actual discovery of this law took him eight years, from 1907 to 1915. It was not an unequivocal consequence of the relativity principle but, rather, demanded a new understanding of how the classical knowledge of energy and momentum conservation and of Newton's law of gravitation was to be incorporated in the new theory; and it demanded a new understanding of space and time.

It is to this new understanding that section 23 is dedicated. Einstein introduces the mental model of a rotating disk considered as an accelerated reference frame. By the general principle of relativity an observer on the disk is entitled to consider the system as being at rest but having a peculiar gravitational field. The observer is equipped with standard clocks and rods; however, using these measuring instruments to coordinate space and time measurements within the rotating frame of reference turns out to be difficult when the implications of special relativity are taken into account. A clock situated at the rim of the rotating disk and a clock sitting at its center will not tick at the same pace because, according to special relativity, they are present in reference frames moving with respect to each other. Pursuing the same thought experiment with rods, Einstein derives the even further-reaching consequence that for the observer on the rotating disk, Euclidean geometry no longer holds, because the behavior of the standard rods also depends on their location of use.

Suppose that an observer in a reference frame at rest with respect to the rotating disk uses standard rods to measure the circumference of the disk. According to special relativity, these rods are shortened with respect to that observer (Lorentz contraction); therefore, more rods are needed, and the length of the circumference exceeds the result of such a measurement for a nonrotating disk. At the same time, the length of the rods used to measure the diameter is unaffected. Therefore the ratio of the circumference to the diameter is greater than π. Thus, Euclidean geometry does not apply to the rotating system.

In a copy of the 10th German edition, found at the Harry Ransom Humanities Research Center at the University of Texas, a sheet of paper is inserted after section 23 with a remark in the handwriting of Ilse Einstein, daughter of Einstein's wife Elsa, who then served as his secretary. The remark refers to the discussion of the rotating disk. It reads: "Note: This interpretation has often been opposed as unconvincing because not only the measuring rods but also the circular disk would suffer tangential contraction. This argument is not cogent because the rotating disk cannot be viewed as a Euclidean rigid body; such body would shatter when brought to rotation, according to the postulated reasons. In reality the disk plays no role in the whole consideration, but only the system—of rods at rest

> relative to each other—which rotates as a totality of radially and tangentially positioned rodlets."[16]

Einstein had used the mental model of a rotating disk as early as 1912 to justify that the new theory of gravitation requires a new framework for space and time. Einstein concludes section 23 with a challenge for the reader: do these implications, which show that the space and time coordinates of special relativity have lost their direct metric meaning, call into question all that has been achieved so far?

GRAVITATION AND GEOMETRY (§§ 24–27)

The following four sections are dedicated to a preparation for what Einstein calls an exact formulation of the general principle of relativity. He first extends his discussion of geometry by discussing the possibility that a non-Euclidean continuum exists in which the familiar measurement processes in Euclidean space no longer work. He then presents the idea of Gaussian coordinates for curved surfaces as a resolution to these difficulties. In a third step he returns to special relativity, illustrating the meaning of Gaussian coordinates for this case. In the last step of his preparation Einstein shows that these generalized coordinates help address the challenge of the meaning of coordinates in a relativistic theory of gravitation.

He begins with a discussion of the Euclidean continuum. The main point of this discussion is that here one can assign a physical meaning to Cartesian coordinates: they label the corners of a grid of identical measuring rods, covering a plane or three-dimensional space. Thus,

16 CPAE vol. 6, Doc. 42, p. 419, n. 49.

Cartesian coordinates are directly connected to distance measurements. This physical meaning of the coordinates will have to be abandoned in the discussion of the non-Euclidean continuum of general relativity.

Einstein then addresses the limits of Euclidean geometry, but he does not immediately turn to the example of a curved two-dimensional surface such as that of a sphere, as is done in most textbook discussions on the role of non-Euclidean geometry in relativity. Instead, Einstein explores a famous example dating to the French mathematician and philosopher Henri Poincaré (1854–1912). In this example a flat surface such as a table is heated in the middle, so that measurement rods expand when they are used in the warmer areas of the surface. Thus, these measurement rods can no longer be used to create the familiar coordinate grid that can normally be imposed on such a flat surface to label its points.

The observer innocently using these measurement rods and confronted with this problem thus has two choices: ascribe the problem either to some physical process affecting the measurement rods, such as the heating of the table, or to the geometry of the table. The problem can easily be resolved if other measurement devices are available that are not subject to deformation by heating. But if all measurement devices behave in the same way, it remains ambiguous whether one should assume that Euclidean geometry nevertheless holds—although it cannot be directly established owing to the intervention of some universal force affecting all measurements—or whether one should admit the possibility of non-Euclidean geometry.

Poincaré believed that the choice of geometry for describing spatial properties is ultimately conventional. Einstein, however, concluded that measurement rods retain their foundational role for defining the geometrical properties of space even if they may be subject to deformations, at least as long as rods of all materials suffer the same deformation. But this is exactly the situation he encountered in the preceding section when discussing the behavior of rods in a rotating frame of reference.

In the following section (25) Einstein introduces Gaussian coordinates as a general means for dealing with continua such as lines, surfaces, and spaces, and also with continua of a higher dimension. He sets aside the mathematical considerations and focuses on the physical meaning of such coordinates. They serve, first, to uniquely label points by sets of numbers in accordance with the dimension of the continuum under consideration, two numbers for a surface, three for a space, and so on. These Gaussian coordinates are the crossing points of sets of curves, covering the continuum in such a way that each point of the continuum is denoted by exactly one crossing point. In contrast with the familiar Cartesian coordinates, these curves no longer have to be straight lines, while neighboring points are still characterized by infinitesimally small coordinate numbers.

In a Cartesian coordinate system, the distance between two points can be calculated by the Pythagorean theorem using a right triangle formed by the distances and the coordinate differences between two points. In a Gaussian coordinate system, the distance is related to the coordinate differences in a more complicated way. In general, the infinitesimal distance between two neighboring points is given as a function of the coordinate differentials (i.e., infinitesimal coordinate differences), where this function depends on the position in the continuum.

Specifically, the square of this infinitesimal distance is given by a sum of terms that generalizes the Pythagorean theorem, at least if it is possible to consider increasingly smaller parts of it as being ever flatter. In this sum, a set of functions multiplies products of the coordinate differentials. The set of these functions forms what is also called the "metric" of the continuum, since it determines how coordinate differences are related to the actual metric distance between two points that can be measured. If the continuum has a Euclidean structure, this sum can always be transformed into a sum of squares of the coordinate differentials, so that the usual Pythagorean theorem is fully recovered. Gaussian coordinates thus make it possible to introduce coordinates and relate them to distance measurements between neighboring points even when no Euclidean continuum is given.

This is the case in Einstein's theory of general relativity where non-Euclidean continua are considered. They do not result from a generalization of ordinary space in the sense of Carl Friedrich Gauss (1777–1855) and Bernhard Riemann (1826–1866) but from a generalization of the space-time continuum of special relativity, as described by the mathematician Hermann Minkowski. The subsequent section (26) therefore takes up the earlier discussion (see section 17) of Minkowski's space-time framework to show that it can be conceived as a "Euclidean" continuum that will have to be generalized later. Einstein begins his discussion with Galilean coordinate systems, representing the privileged inertial frames of classical physics and special relativity. In special relativity, events observed in different reference systems are related by Lorentz transformations, characterized by Einstein as the expression of the universal validity of the law of light propagation in such systems. He then shows that these transformations satisfy the condition that the four-dimensional infinitesimal distance is the same in all Galilean reference frames. The constancy of the speed of light can thus simply be used as a factor multiplying the time coordinate, and the square of this infinitesimal distance can be rewritten as a sum of squares of the coordinate differentials, in agreement with the usual Pythagorean formula and with the notion that these coordinate differentials themselves have a direct physical meaning.

Einstein is now finally in the position to summarize his reasoning till this point to explain how the introduction of Gaussian coordinates overcomes the difficulties mentioned and allows the space-time of general relativity to be described as a non-Euclidean continuum. Earlier, he showed that in general relativity the speed of light can no longer be constant if a gravitational field is present and that such a field also makes it impossible to use physically meaningful coordinates, as in the case of special relativity. He now refers to the example of the heated table for which it was impossible to use measurement rods for constructing a Euclidean grid of coordinates. This example motivates his assertion that the problem of gravity is of the same nature and can be solved by the same means, the introduction of Gaussian coordinates. Einstein then proceeds to show how these coordinates can be employed to label events in space and time without presupposing that the underlying geometry is Euclidean.

But divesting coordinates of their meaning as direct measures of space and time naturally raises the question, what is their meaning at all? This query is related to even more profound issues that Einstein does not mention in this popular exposition but that nevertheless shaped the answer he offers to this question in this last preparatory section before turning to the formulation of the general principle of relativity. In fact, before he completed his theory in 1915 Einstein had developed a preliminary version characterized by a restricted choice of admissible coordinates. He had tried to justify this theory by an argument—which later became famous as the "hole argument"—that assumes single points in the continuum can be labeled and ascribed a meaning independent of physical events associated with them. After discarding the preliminary version and completing the theory of general relativity, Einstein learned, probably from his discussions with the philosopher Moritz Schlick, how to discard the hole argument. He realized in particular that events in space and time can be identified only by coincidences of real physical processes, such as the encounter of two particles, represented by the crossing of two lines in space-time and describing the motion of the particles. From then on he insisted, as he does in this popular exposition, on the meaning of space-time coincidences and their characterization by Gaussian coordinates as "the only actual evidence of a time-space nature with which we meet in physical statements." (p. 111)

GRAVITATION AND GENERAL RELATIVITY (§§ 28–29)

In the last two sections of his booklet, Einstein reformulates the general relativity principle and proposes to solve the problem of gravitation on its basis. The earlier formulation made use of rigid reference bodies, which are now replaced by Gaussian coordinate systems, because rigid bodies with Euclidean properties are, in general, no longer available. To give these Gaussian coordinate systems a physically intuitive meaning, Einstein introduces the idea of a "reference mollusk," which is conceived as a nonrigid body in arbitrary motion equipped with clocks at each of its points. He gives three formulations of the general principle of relativity:

- All Gaussian coordinate systems are equivalent for the formulation of the general laws of nature.
- The equations expressing these laws are transformed into equations of the same form under arbitrary substitutions of Gaussian coordinates.
- The laws of nature are independent of the reference mollusk chosen to formulate them.

It turned out that it is possible also to formulate the Newtonian theory of gravitation in such a coordinate-independent way, so it is disputable whether Einstein's quest for a generalization of the relativity principle singles out his general theory of relativity. But this theory is not only more natural when formulated in this way, it is also characterized by being background-independent, that is, its geometrical structures are subject to dynamical laws. There are hence in general no privileged coordinate systems like the inertial systems of classical physics, which are the expression of a specific symmetry of its space-time continuum. In any case, Einstein's general principle of relativity suggested conceiving of gravitation and inertia as two aspects of the same fundamental field that may manifest itself differently depending on the state of motion of the observer but nevertheless obeys the same laws. It was this insight and its formulation in terms of Gaussian geometry that became the basis of the new theory of gravitation put forward by general relativity.

Basic insights into the nature of this generalized gravitational field can indeed be obtained by considering physical effects in an arbitrarily moving reference frame. In the last section (29), Einstein thus takes up his considerations based on the equivalence principle, generalizing them and reformulating them in terms of Gaussian coordinate systems. He again starts from a Galilean reference frame in which no gravitational field exists and then treats gravitational fields observed after transformation to an accelerated reference frame with the help of special relativity, which explains the behavior of rods, clocks, and material points in such special fields. The crucial point is now to assume that the laws thus found also hold in more general gravitational fields. It is immediately found, for instance, that the law of motion of free particles in a gravitational field can be described by a geodetic line representing the generalization of inertial motion to a curved space-time framework. This procedure reflects Einstein's own historical trajectory that in 1912 first led him to establish this and other insights about physical processes in an arbitrary gravitational field.

The subsequent problem was to find the laws governing the gravitational field itself. To find them took Einstein another three years, but his popular exposition hardly reflects the convoluted pathway to the formulation of the field equations of general relativity. To maintain a level of description accessible to the nonprofessional reader, Einstein refrains from describing the field equation for the gravitational field and makes no attempt to explain the concept of space-time curvature and of the geodetic line, which defines the trajectory of a particle moving in a gravitational field. In contrast, in his treatment of special relativity he presents the explicit mathematical form of the Lorentz transformation and demonstrates its derivation in an appendix. Here Einstein simply emphasizes the principles guiding the derivation of the field equation.

The first step is to characterize the special case of gravitational fields generated by accelerated motion in a way that is independent of the reference frame. This special case corresponds to the situation in which no material sources of the gravitational field are present. The more general case in which such sources are present can be found by imposing three conditions: (1) the field equation must also satisfy the general principle of relativity, (2) the matter that is the source of the gravitational field must be represented by its energy content, and (3) the field equation must satisfy the principle of the conservation of energy and momentum. Indeed, these conditions, together with the condition that the general theory should cover the familiar Newtonian theory as a limiting case, guided Einstein's heuristics during the three years he searched for the gravitational field equation.

After briefly mentioning how to find the laws for any physical process within a gravitational field, such as the laws of electromagnetism, Einstein concludes with a brief discussion of the achievement of the theory of general relativity. He first extolls its beauty and its superiority with respect to classical mechanics and then turns to discuss the great success of the theory in explaining "a result of observation in astronomy, against which classical mechanics is powerless." (p. 119) He is referring here to the perihelion motion of planet Mercury that had been observed by astronomers, constituting a well-known if somewhat hidden puzzle of classical celestial mechanics.

The explanation of this motion played an important role in the evolution of the theory of general relativity. In December 1907, when Einstein was taking his initial steps toward a relativistic theory of gravitation and even before he had any kind of a theory, he already realized that it could provide a solution to the long-standing problem. He wrote to his friend Conrad Habicht: "At the moment I am working on a relativistic analysis of the law of gravitation by means of which I hope to explain the still unexplained secular changes in the perihelion of Mercury."[17]

Einstein also mentions two other phenomena that could be deduced from the theory—light deflection in a gravitational field, and the gravitational redshift. At the time he wrote this booklet, only the Mercury perihelion motion was known. He therefore discusses only that phenomenon here. After the gravitational bending of light was confirmed in 1919, Einstein added an appendix (appendix 3) in which he discussed in detail the three classical tests of general relativity.

Referring to the gravitational bending of light and to the gravitational redshift, the last sentence in the first German edition was: "I have no doubt that these consequences of the theory will be confirmed." In the English edition reprinted here, this sentence reads: "These two deductions from the theory have both been confirmed." (p. 121) This statement was premature regarding the gravitational redshift. We shall return to this point in the commentary on appendix 3.

THE CHALLENGE OF COSMOLOGY (§§ 30–32 AND APPENDIX 4)

Einstein's treatment of cosmology in his booklet was added in 1918. It reflects a period in which cosmology was still a field of speculation or, rather, of extrapolations of the consequences of physical theories, allowing their consistency to be probed. After completing his general theory of relativity,

17 Einstein to Conrad Habicht, December 24, 1907, CPAE vol. 5, Doc. 69, p. 47.

Einstein immediately realized that it was relevant to the universe as a whole. Although none of the modern observations were available, the theory imposed some mathematical constraints that could be applied to examining the structure of the universe. Einstein's discussion of this subject may be viewed as the genesis of modern cosmology.

In the first section dedicated to this theme Einstein returns to considerations of the cosmological implications of Newton's theory by the German astronomer Hugo von Seeliger (1849–1924), who had shown that Newton's law leads to a rather unsatisfactory picture of the cosmos. This representation turns out to be incompatible with the intuitive picture of an infinite universe filled with stars, which on a large scale are homogeneously distributed everywhere.

In the second section Einstein returns to another line of speculation about the constitution of the world involving non-Euclidean geometry, as discussed by Bernhard Riemann, Hermann Helmholtz (1821–1894), and Henri Poincaré. He examines in some detail the case of a finite world without boundaries, which he illustrates with two-dimensional beings living on the surface of a sphere and then extends it to three-dimensional space. Such a "finite" but "unbounded" universe not only would resolve the difficulty raised by Seeliger but also promised to finally provide Einstein with a satisfactory explanation of inertial effects, which—following an idea born out of his reading of Ernst Mach[18]—he suspected were caused by the presence of distant masses.

18 Mach, Ernst. *The Science of Mechanics: A Critical and Historical Account of Its Development* (Lasalle, IL: Open Court, 1960).

The last section of the booklet presents Einstein's static universe, first published in his 1917 paper "Cosmological Considerations in the General Theory of Relativity."[19] Einstein's seemingly natural assumption of a static universe appeared to fulfill his hopes of explaining inertia—in Mach's sense—as an interaction with the stars that fill such a universe. However, Einstein's original field equations do not include any solutions representing a static spatially homogeneous mass distribution. To solve this problem, he introduced a new constant, called the *cosmological constant*, which essentially represents a small repulsive force acting on a large scale. In Einstein's model the nonvanishing matter density is directly related to the radius of the world.

In the 14th English edition, published in 1946, Einstein added, as a supplement to section 32, an appendix dealing with the expansion of the universe. He mentions the discovery of dynamical solutions to the field equations of general relativity by the Russian mathematician Alexander Friedmann (1888–1925) in 1922 and the observations of the Doppler redshift of distant galaxies by the U.S. astronomer Edwin Hubble (1889–1953) in 1929. The latter showed that galaxies are moving away from us and suggested the abandonment of the static picture of the universe in favor of an expanding universe, which later gave rise to what came to be called the *Big Bang model*. Einstein's cosmological considerations end on a skeptical note because at the time he was writing—on the assumption of an expanding universe—the universe seemed to be younger than the earliest known cosmic objects.

19 CPAE vol. 6, Doc. 43, pp. 421–432.

The evolution of Einstein's ideas on the universe as a whole—described by him in the third part of the booklet and in the appendix—merits a few additional remarks. Today, general relativity is considered to be the theoretical basis of observational cosmology. But cosmology played hardly any role in the genesis of the theory. Epistemological considerations and, in particular, Mach's criticism of Newton's absolute space were far more important to Einstein's thinking. He followed Mach's idea to conceive inertial effects occurring in accelerated frames of reference not as a consequence of motion with respect to absolute space but as owing to an interaction with distant masses. This idea played a key role in Einstein's heuristics. General relativity was supposed to explain all inertial effects by such interactions. When it later turned out, to Einstein's surprise, that the theory did not comply with this expectation, he decided, in 1917, as mentioned earlier, to modify the theory by introducing an extra term involving the cosmological constant. This allowed Einstein to formulate the solution discussed in the book, a static and closed universe with a finite matter-density in agreement with his Machian expectations. When Friedmann discovered, in 1922, an expanding-universe solution of the original field equation, Einstein initially dismissed it as a calculation error. And even when he recognized that this was not the case, Einstein denied any physical relevance to such a solution. He was equally dismissive when he learned about the rediscovery of expanding- universe solutions five years later by the Belgian Jesuit priest and astronomer Georges Lemaître (1894–1966). It was only when Hubble's observations supported such a view that Einstein abandoned his original proposal of a static universe and with it the cosmological constant he had introduced to save the Machian interpretation of general relativity.

Today, the description of the expanding universe has taken another turn. Modern observations make it possible to measure not only the present rate of expansion but also how the rate has changed over time since the Big Bang. To account for these findings, general relativity again makes use of a cosmological term.

THE RELATION BETWEEN THEORY AND EXPERIMENT (APPENDIX 3)

Einstein begins this appendix with a discussion of the evolution of a scientific theory, which is evidently more complex than a straightforward inductive process based on empirical observations. He emphasizes the role of intuition and deductive thinking in formulating the principles that constitute a theory. Such a process may lead to different theories that are compatible with the empirical facts. It is important then to find new predictions that may distinguish among them.

Einstein points out that conceptually speaking, the theory of general relativity and Newtonian mechanics are worlds apart, yet they are distinguished by minor empirical differences. This thinking has now changed significantly, because on the cosmological scale there are major differences in the predictions of these two theories. At Einstein's time, however, predictions of rather minor effects that could be confirmed or rejected by direct observations were very important in distinguishing it from the predictions of Newtonian theory and also from alternative relativistic theories of gravitation.

One alternative theory of note was the gravitational theory of the Finnish theoretical physicist Gunnar Nordström (1881–1923), published in 1912. Nordström's theory was based on a single scalar gravitational potential and embedded in the special theory of relativity. Einstein discussed Nordström's theory extensively in a lecture he gave in Vienna in September 1913, "On the Present State of the Problem of Gravitation." He concluded that, "in sum, we can say that Nordström's scalar theory, which adheres to the postulate of the constancy of the velocity of light, satisfies all of the conditions that can be imposed on a theory of gravitation given the current state of empirical knowledge."[20] Indeed, Einstein considered this to be the only viable alternative to his own theory. In Nordström's theory, there is no deflection of light by a gravitational field. It was hoped that the planned astronomical observations during the solar eclipse in 1914 would verify this. Although no observations could be realized at that time, the verdict about the shortcomings of Nordström's theory became clear even before the next solar eclipse in 1919. A calculation of the perihelion shift of Mercury based on Nordström's theory predicts a retrogression, while Einstein's theory predicts the observed progression of 43" (arc seconds).

Kepler's laws, which state that the planets move around the sun in elliptic orbits, can be derived from Newton's theory of gravitation. If there were only one planet in the solar system, then the position of the perihelion of the orbit would be fixed in space. However, owing to the influence of the other planets, there is a slow precession of the perihelion. Astronomers discovered that the orbit of the

20 CPAE vol. 4, Doc. 17, p. 207.

planet Mercury around the sun rotates, as seen from Earth, by 5600", or 1.55° (degrees), in 100 years. Most of this rotation could be explained by forces exerted by the other planets, but 43" was unexplained. In 1859, the French astronomer Urbain Le Verrier (1811–1877) discovered Mercury's precession. Attempts to explain it were unsuccessful until Einstein's general theory of relativity. Le Verrier was the first to report that the slow precession of Mercury's orbit around the Sun could not be completely explained by Newtonian mechanics and perturbations by the known planets.

Einstein began to work on the observational consequences of the theory of general relativity in 1911. He predicted two effects that could provide decisive tests of the theory. The first was that light would be deflected in a gravitational field. In 1913, he wrote to the astronomer George Hale (1868–1938) asking for his advice on the possibility of measuring the deflection of light in the vicinity of the solar rim.[21] Hale's reply was that it would be possible to detect this effect only during a solar eclipse.[22] In 1914, after the breakout of World War I, a German expedition was planned to observe this effect during a solar eclipse in the Ukraine, but the expedition was interned for a brief period by the Russian authorities and missed the eclipse.

Einstein's prediction was confirmed by astronomical observations during the solar eclipse of 1919, performed by a British expedition headed by the astronomer Arthur Eddington (1882–1944), and Einstein became a world celebrity overnight. A bending of light by gravity had been

21 Einstein to George Hale, October 14, 1913, CPAE vol. 5, Doc. 477, pp. 356–357.

22 George Hale to Albert Einstein, November 8, 1913, CPAE vol. 5, Doc. 483, p. 361.

inferred from Newton's particle theory of light but fell into oblivion after the triumph of the wave theory of light. The predicted angle of deflection was smaller by a factor of 2 than the value following from Einstein's theory. It was therefore not just the phenomenon itself but also the measured angle that caused the *Times* of London to publish a three-deck banner headline on November 7, 1919: "Revolution in Science—New Theory of the Universe—Newtonian Ideas Overthrown."

The second effect Einstein predicted is the change in the color of light in a gravitational field—the so-called gravitational redshift—which is a consequence of the slowing down of the rate of clocks in the neighborhood of a massive body. An atom emitting light may be viewed as a clock. The slowing down of such "atomic clocks" in a gravitational field means that the frequency of oscillations associated with the motion of the electrons in the atom, and hence of the emitted light, is reduced. The "color" of light of lower frequency is shifted toward the red end of the spectrum of light.

When Einstein wrote this appendix, the empirical evidence for this phenomenon was inconclusive, and he stated that if this prediction could not be confirmed, "then the general theory of relativity will be untenable." (p. 152) In the English edition reprinted here, a footnote by the translator on the same page states that it was "definitely established by Adams in 1924, by observations on the dense companion of Sirius." It turned out that this conclusion was also premature, and the redshift was not definitively confirmed until the late 1950s, when R. V. Pound and G. A. Rebka compared the frequency of light emitted by a specific isotope of cobalt at the bottom and at the top of a 22.5 m-high tower. They measured a frequency shift equivalent to the slowing down of a clock at the bottom of the tower by about one second in 30 million years. This experiment marked the beginning of an era of high-precision measurements to test the general theory of relativity. Nowadays, the effect of gravitation on the rate of clocks has to be taken into account in the timing of global positioning system (GPS) technology.

THE CHANGING CONCEPT OF SPACE (APPENDIX 5)

The fifth appendix was added to the 16th German and the 15th English edition of Einstein's book in 1954, the last to appear in his lifetime. In the preface, dated 9 June 1952, Einstein wrote: "I wished to show that space-time is not necessarily something to which one can ascribe a separate existence, independently of the actual objects of physical reality. Physical objects are not in space, but these objects are physically extended. In this way the concept "empty space" loses its meaning." In this appendix he reviews the evolution of the concept of space from physical intuition to the implications of general relativity. At the same time, he presents a concise summary of the development of his own thinking on a key concept of physics that preoccupied him throughout his lifetime.

His central question concerns the physical reality of space. His conclusion is that the introduction of the field concept has made it possible to conceive of space no longer as a preexisting stage for physical events but as a dynamical part of the physical reality described by a field. The text displays remarkable similarities to Einstein's *Autobiographical Notes* and was evidently written with the future development of physics in mind. It may be considered Einstein's epistemological legacy regarding the concept of space.

The other four appendixes are supplements to particular issues or chapters in the main body of the text. In three cases this purpose is even specified explicitly in the appendix title. However, this appendix stands alone as an independent essay in which the ideas leading from classical physics to special and general relativity are reviewed in the context of the development of the concept of space. It is possible that Einstein decided to include it in his booklet when he realized its undiminishing popularity, 35 years after its first publication.

Notions of Space: Prescientific, Euclidean, Cartesian, Newtonian

Einstein's point of departure is the remarkable fact that the concept of acceleration plays a fundamental role in Newton's law of motion, which seems to ascribe physical reality not just to space but also to its state of motion. Einstein then explains his uneasiness about this by referring to the philosophical position of René Descartes (1596–1650) that empty space cannot exist, because extension must always be related to bodies. Strange as this position may appear at first glance, it is in some sense redeemed by general relativity, as Einstein claims. He then pursues the psychological origins of the concept of space and finds it in the mental model of a container or box in which objects can be placed according to the laws of Euclidean geometry, which are thus an expression of experiences in the real world and not given a priori. While the box model may be intuitively plausible, using it for dealing with motion in physics makes it necessary to extend the model by introducing the idea of inertial reference frames—corresponding to infinitely many such spaces in motion relative to each other—which is no longer intuitively evident.

Einstein next turns to the psychological origins of time, finding it in the distinction between sensual experiences and memory and ascribing it to an ordering intervention of reason. This subjective concept of time may be extended to an objective concept by including the experience of other persons reacting to the same event, which only through their sharing of experience takes on the meaning of an objective event. Einstein emphasizes that the temporal order of experiences does not imply the temporal order of such objective events. To establish the latter requires taking into account their spatial position.

The concept of space has a complex architecture, with experiences involving the placement of objects constituting a fundamental level. Even more basic is the concept of the object itself, as Einstein stresses. His aim is to clarify the empirical aspects of such concepts as space, time, event, and material object to be able to adjust them to new experiences if necessary. This necessity is, of course, precisely what led to the overturn of classical concepts in Einstein's relativity revolution. He acknowledges the role of David Hume and Ernst Mach in having initiated such a critical reflection of fundamental concepts, a role mentioned earlier in the context of Einstein's crucial steps toward his special theory of relativity.

Einstein then carefully addresses the transition from prescientific to the scientific concepts emerging from them. Again he stresses the role of experience in shaping these concepts, in particular the dependency of Euclidean geometry on experiences with the placement of rigid bodies, which brings him back to the body concept he examined earlier in the context of intuitive physics. However, at the level of science, bodies cannot be considered in isolation from the rest of physics, and physics in turn depends on geometry. Einstein thus concludes that because of this interdependence, the empirical content of geometry can be identified only in the context of physics as a whole. This limitation is particularly evident in the case of microphysics, where direct spatial measurements are impossible. Despite these difficulties, the concept of space with its intuitive foundation and its important role in all science seemed indispensable. Only Mach attempted, as Einstein points out, to replace space with the set of all distances between material points, to find an alternative to Newton's explanation of inertia in relation to absolute space.

The Concept of Field *and Its Emancipation from Its Mechanical Roots*

In the next step Einstein reviews the origin of the field concept and its impact on the understanding of space. He begins by reminding the reader of the double role of space and time in classical physics: as a stage on which physical events are played out and as the set of inertial systems for explaining physical events. In classical physics, space and time are independent from matter, which is conceived of as fundamentally comprising material points whose motion is at the basis of all physical processes. If these material points were to disappear, space and time would still remain.

This conception was gradually undermined by the introduction of the field concept, which was first used in cases when matter could be treated as a continuum to express certain of its properties as functions of space and time. But when such a description turned out also to describe light as a wave phenomenon, a paradoxical situation emerged, since this entailed the existence of an all-pervading hypothetical material medium (the ether) carrying those light waves and the fields corresponding to them. The fundamental role of the field concept became even clearer with the formulation of the laws of electromagnetism by Faraday and Maxwell and the realization that light waves are electromagnetic waves. Einstein finds this gradual emancipation of the field concept from its material substratum psychologically most remarkable and underlines the fact that a concept originally born within classical mechanics eventually transcended its framework to become a foundational concept in its own right, without needing the support of an ether.

This liberation of the field concept from its mechanical roots was, however, not a smooth process, as it meant breaking with the idea, seemingly self-evident in the context of classical physics, that where there are waves there must be a medium carrying them. Prior to this break numerous attempts were made to explore the consequences of this idea, in particular with regard to the nature and the state of motion of this hypothetical medium. This exploration eventually led to a formulation of electromagnetic theory by the Dutch physicist H. A. Lorentz in which the failure to actually detect motion with regard to the hypothetical ether was reconciled with its foundational role as the carrier of electromagnetic fields. On its face, this theory could appear to be a satisfactory solution of the problem that the ether should exist and remain immobile while motion with respect to it should be undetectable (see section 16 and the related commentary).

But when the young Einstein struggled with this problem, he refrained from restricting his vision to just one subfield of physics and tried to consider its entire conceptual foundation. Indeed, from the perspective of mechanics, Lorentz's solution was paradoxical. The assumption of an ether at rest was in contradiction with the claim of mechanics that there is no privileged reference frame at absolute rest and that, according to the principle of relativity, the laws of physics should be the same in all inertial frames. The undetectability of motion with regard to the ether—confirmed by experiments that with some effort could be explained by Lorentz's theory—was in agreement with the principle of relativity but the latter was not part of that theory. From the perspective of mechanics there was thus a mismatch between the experiments described by the theory of electromagnetism and its conceptual foundation.

Space, Time, and Field in Special Relativity

Einstein's special theory of relativity, formulated in 1905, offered a solution to precisely this conflict. The title of his original paper, "On the Electrodynamics of Moving Bodies," makes it clear that he was dealing with a borderline problem between electromagnetism and mechanics. About half this paper deals with the kinematics of motion of material bodies, and Einstein arrived at his solution by preserving key insights of both electromagnetism and mechanics. From the Maxwell-Lorentz theory of electromagnetism he took the principle that the velocity of light in vacuum is a constant, while from classical mechanics he retained the relativity principle and extended it to electromagnetic phenomena.

However, these two principles cannot be reconciled within the context of the classical understanding of the concepts of space and time. This motivated the young Einstein to reflect on the nature of these concepts in light of the epistemological considerations discussed earlier in this appendix. A revision was required precisely at this deeper conceptual level, introducing a new concept of simultaneity and a new transformation law for the relation between space and time measurements in inertial frames that are in relative motion with respect to each other. Einstein did not have to invent this new transformation law—he was able to learn it from Lorentz's theory, which had introduced space and time transformations as auxiliary devices to cope with the undetectability of motion with respect to the ether. The formulation of the theory of special relativity was largely based on using the technical framework of the earlier theory as a structure for introducing new foundational concepts.

After this brief review of the genesis of special relativity, Einstein returns to the problem of space. He addresses the popular idea that relativity theory has supposedly revealed the four-dimensional character of the world, making it clear that the specification of events by three space coordinates and one time coordinate was already part of classical physics. At this point, without explicitly introducing Minkowski's four-dimensional framework, he emphasizes that it is the nature of this space-time continuum with regard to the separability of space and time that is different according to relativity theory. Here the space-time continuum can no longer be objectively subdivided into subsets containing all simultaneous events, since the notion of simultaneity has become dependent on the reference frame chosen. He concludes this discussion with the implication of special relativity for the ether concept. Since the theory has established the equal validity of all inertial frames, there cannot be an ether at rest. As a further consequence, the concept of field is freed from any material carrier.

In a last step prior to addressing the concept of space according to general relativity, Einstein reviews the commonalities between the space-time of special relativity and the space of classical physics. In both, inertial frames of reference play a privileged role, and in both they constitute an independent stage for physical events, which continue to exist even when all matter and fields disappear. Since in Minkowski's space-time, physical events and processes can be described as four-dimensional geometrical entities, physical reality is more naturally conceived of as *being* rather than as *becoming*. For instance, the life history of a particle is a line and not a flow process. Minkowski's four-dimensional space-time is a framework existing independently from the physics taking place within it. The concept of space-time in special relativity therefore does not comply with Descartes's denial that space can exist independently from matter.

With his exhaustive survey of the evolution of the concept of space from physical intuition via classical mechanics to Minkowski's space-time, Einstein thus created tension by presenting the question of the primacy of matter for the concept of space as the result of a long-standing conceptual history. When he began to work on general relativity in 1907 he was essentially driven by the same concern: whether Mach's idea of explaining inertia by the interaction of masses would make it possible to discard Newton's absolute space. But in spite of the success of general relativity, Einstein had to learn that this program had its difficulties, in particular because general relativity also allowed for inertial effects even in the absence of matter. He now addresses the same challenge from a point of view that centers on the primacy of the field concept.

The Concept of Space in the General Theory of Relativity

Einstein's heuristic starting point for the formulation of general relativity is the equivalence principle, relating an inertial system to a reference frame in uniformly accelerated motion with respect to it. These may be considered equally valid reference frames if a homogeneous gravitational field exists in the second frame that balances the acceleration. In an accelerated frame of reference all masses evidently fall with the same acceleration independent of their constitution, so that here the equality of inertial and gravitational mass is automatically guaranteed. If this equivalence principle is extended to arbitrary relative motions, inertial frames lose their privileged role. A theory based on the equivalence principle thus brings inertia and gravitation together in a single framework and naturally incorporates the equality of inertial and gravitational mass.

The question that Einstein addresses next is that of the nature and range of allowable transformations of space and time coordinates, generalizing the Lorentz transformations. Inertial frames are characterized by the fact that coordinates have a simple and direct relation to space and time measurements. In transformations to accelerated reference frames this relation is lost, as discussed in the commentary on section 10. But once the relation is lost, it becomes natural to assume that all possible continuous space-time transformations are legitimate, since coordinates no longer have to express metric properties of space; namely, they are no longer directly related to the measurement of distances. This consequence implies, according to Einstein, the general principle of relativity and the requirement that the laws of nature should remain the same (more precisely, remain covariant) under such arbitrary coordinate transformations. Einstein claims that this requirement, together with that of logical simplicity, restricts the admissible laws more stringently than does special relativity. He next sketches, without any details, the path by which the field equation of general relativity can be derived from such general considerations. This path does not correspond to Einstein's original, tortured route to general relativity and very much reflects his later conviction that physical theories can be found by such criteria of logical and mathematical simplicity. It is particularly remarkable that Einstein was evidently convinced that the pathway to general relativity is clearly laid out once the nature of the field as an independent or irreducible concept is seen.

This consideration also underlies the conclusion of his essay, which explains how the transition to general relativity modifies the concept of space. Space no longer has a "special" existence independent of what fills it. Space—or more precisely, space-time—is described with the help of a mathematical concept called a *metric* that determines the relation between coordinates and the measurement of distances. Einstein now considers this concept as simultaneously representing a field. This description also includes the Minkowski space-time of special relativity as a special case representing a particularly simple case of a metric. If this "field" is taken away, nothing remains, not even space. Of course, this argument is possible only if the inertial forces occurring in such a space-time are considered as consequences of a generalized gravitational field. The reasoning is based on the double nature of the metric, simultaneously characterizing the gravitational field and the geometry of space. For Einstein, this double nature is, in turn, a consequence of the field concept taken together with the general principle of relativity. This double nature accordingly leads to the conclusion that "space-time does not claim existence on its own, but only as a structural quality of the field" and that the modern interpretation of Descartes's idea is "there exists no space 'empty of field'." (pp. 176–77) This last sentence appears in the English edition but is missing in the German; however, the idea is clearly expressed in the preface to both editions. This completes Einstein's goal to unify the concepts of matter and space. He is pleased to acknowledge that he was preceded in this quest by a great thinker like René Descartes.

Einstein ends his essay with two remarks concerning the contemporary state of physics; the first relates to his own attempts at creating a unified field theory, and the second refers to the foundations of quantum theory of which he had become skeptical. His interpretation of the success of general relativity as a result of pursuing the far-reaching consequences of the introduction of the field concept suggested to Einstein that other physical forces also may be integrated into this framework, conceiving of them as a generalization of the law for the gravitational field. However, as he notes, such a generalization lacks the natural starting point that was afforded the search for general relativity by the special case of Minkowski space-time. Nevertheless, even in this popular exposition, Einstein briefly points to the version of such a generalized field theory that he had come to believe in. The essay ends with an equally succinct criticism not so much of quantum theory itself but of the belief of the majority of contemporary physicists that the physically real can no longer be described by a field but, rather, only by statistically measured results. Einstein, in contrast, is convinced that the avenue indicated by the field concept should be further pursued.

These concluding remarks are in line with our suggestion to perceive this chapter as Einstein's epistemological legacy concerning the concept of space.

A History and Survey of Foreign-Language Editions

The success of Einstein's booklet in Germany was followed by a similar, unprecedented success in other countries, and within a short period it was translated into many languages. Einstein was personally involved in this process, in most cases through his German publisher, Vieweg, and in other cases dealt directly with foreign publishers. He approved translation initiatives, corresponded with the translators, and negotiated their and his compensation rates.

The first English translation was published in 1920, followed by those in French (1921), Italian (1921), Japanese (1921), Polish (1921), Russian (1921), Spanish (1921), Hungarian (1921), Chinese (1922), Czech (1923), and Hebrew (1928). Some of these first editions were quickly followed by new editions. In 1921, there were also inquiries, which did not materialize, about translations into Croatian, Latvian, Estonian, Ukrainian, and Yiddish. Einstein also gave his permission to Vieweg to produce the booklet in Braille script, but it is not known whether this actually happened.

SAMMLUNG VIEWEG

TAGESFRAGEN AUS DEN GEBIETEN
DER NATURWISSENSCHAFTEN
UND DER TECHNIK

Heft 38

Über die spezielle und die allgemeine Relativitätstheorie

Gemeinverständlich

Von

A. Einstein

VIEWEG & SOHN BRAUNSCHWEIG

Über die spezielle und die allgemeine Relativitätstheorie

(Gemeinverständlich)

Von

A. EINSTEIN

Mit 3 Figuren

Braunschweig
Druck und Verlag von Friedr. Vieweg & Sohn
1917

Many other foreign-language editions appeared from the 1950s onward: Arabic, Armenian, Bulgarian, Croatian, Dutch, Greek, Hungarian, Icelandic, Norwegian, Portuguese, Romanian, Serbian, Swedish, and Turkish.

We shall discuss some of the foreign-language editions of the 1920s and point out their role in the spread of the theory of relativity and its public understanding in the respective countries.[1]

THE ENGLISH TRANSLATION

The sensational headline in the *Times* of London on November 7, 1919, "Revolution in Science—New Theory of the Universe—Newtonian Ideas Overthrown," announced the confirmation of one of the predictions of the theory of general relativity, the bending of light in the gravitational field of the sun. About two weeks later, Robert W. Lawson, a lecturer of physics at the University of Sheffield, wrote to Einstein to suggest he publish, in *Nature*, an article on the theory of relativity and gravitation that he would translate into English. The article was to be short (3000 words) and understandable to nonmathematical readers. Lawson congratulated Einstein on the observation of the gravitational bending of light and told him that for several weeks people had been talking about nothing else.

Einstein initially agreed to write the article but later decided that what he had written was not appropriate for publication in *Nature*. In the meantime, Lawson proposed to translate the booklet on the "special and general theory." He believed that a generally understandable account

1 A list of works on the reception of Einstein's relativity theories is given in the Further Reading section.

RELATIVITY

THE SPECIAL AND GENERAL THEORY

BY

ALBERT EINSTEIN, Ph.D.

PROFESSOR OF PHYSICS IN THE UNIVERSITY OF BERLIN

TRANSLATED BY

ROBERT W. LAWSON, M.Sc.

UNIVERSITY OF SHEFFIELD

NEW YORK

HENRY HOLT AND COMPANY

1920

of this subject would be in great demand. Lawson asked Einstein to write a special introduction to the English edition, to include his portrait, to send a brief life story, and to write an addendum on the empirical confirmation of the theory. Apart from writing a special introduction, Einstein fulfilled all these requests. Only later did he agree to comply with requests to write special introductions in a very few cases. For the portrait, he sent a copper etching by Hermann Struck, which was later included in several foreign-language editions.

Before the publication of the English edition, Lawson wrote to Einstein about an amusing letter he had received from the publisher Methuen requesting that in the promotional material for the book, he should make "the description of its contents as intelligible as possible to the ordinary man. Our travellers tell us that there is complete ignorance in the public mind as to what Relativity means. A good many people seem to think that the book deals with the relations between the sexes. Perhaps you would explain the meaning of the word and say something about the epoch-making character of the book and how Einstein's discovery affects Newton's law. Most people have heard of Newton and his apple and that will give some kind of a clue."[2]

The first edition appeared in England in August 1920 and was followed by four new editions within two years. Lawson's translation was published simultaneously in the United States with similar success.

2 Robert Lawson to Einstein, February 22, 1920, CPAE vol. 9, Doc. 326, p. 273.

THE FRENCH TRANSLATION

In February 1920, Jeanne Rouvière, a student of the prominent French mathematician and active politician Émile Borel, asked Einstein for permission to translate his booklet on the special and general theory of relativity into French. She wrote that the translation would be made under the guidance of Borel, who would also write a preface to introduce the booklet to the French readers. Einstein agreed, and the first French edition of the booklet appeared in 1921.

A few months after the first contact with Mme. Rouvière, Einstein was approached by his friend Maurice Solovine. During his years in Bern (1902–1908) Einstein had formed a reading club with two friends, Maurice Solovine and Conrad Habicht, which they called "Olympia Academy." Together, they read classical literature on science and philosophy. Einstein responded, saying that he had already granted permission for this translation but authorized Solovine to translate several other works into French. Some time later, Einstein received comments about certain errata in the French translation, and his secretary instructed the publisher to involve Solovine in the necessary corrections and modifications before the next edition.

Ultimately, Einstein demanded a new French translation by Solovine, who was grateful and very happy to undertake the project. Solovine suggested calling it a new augmented edition by making minor textual changes and adding one of Einstein's earlier papers from 1911 on the influence of gravitation on light propagation. He hoped that this would appease Jeanne Rouvière. Einstein objected to the inclusion of the 1911 paper in the new edition because it gave

ACTUALITÉS SCIENTIFIQUES

LA THÉORIE DE LA RELATIVITÉ RESTREINTE ET GÉNÉRALISÉE

(MISE A LA PORTÉE DE TOUT LE MONDE)

PAR

A. EINSTEIN

TRADUIT D'APRÈS LA DIXIÈME ÉDITION ALLEMANDE

Par Mlle J. ROUVIÈRE,
Licenciée ès sciences mathématiques.

Avec une Préface de M. Émile Borel.

PARIS,
GAUTHIER-VILLARS ET Cie, ÉDITEURS
LIBRAIRES DU BUREAU DES LONGITUDES, DE L'ÉCOLE POLYTECHNIQUE
55, Quai des Grands-Augustins.

1921

the predicted angle of the gravitational bending of light as only half of the correct value, which he first calculated after completing his theory of general relativity. Solovine then suggested that Einstein expand the first chapter recounting his student experiences with Euclidean geometry. Solovine's translation was finally published in 1923 with no significant changes and with no introductory remarks.

A key figure in the dissemination of the theory of relativity in France was the physicist Paul Langevin, who had encountered Einstein's work as early as 1906 and had subsequently become an ardent supporter. Between 1910 and 1911, he gave a course on relativity at the Collège de France. He also shared Einstein's antimilitarist position and in 1922 invited him to give lectures in Paris. Before that visit, in the early 1920s, Einstein's theories of relativity had generated little interest among physicists (particularly the general theory of relativity). The greatest interest in the new theory could be found among mathematicians, and also among engineers and philosophers. It was therefore no coincidence that the initiative to translate Einstein's booklet came from a mathematician.

The discussions around the new theories ranged from skepticism to objections. In January 1922, two months before Einstein's visit to Paris, a debate on the theory of relativity took place at the Academy of Science. One of the most prominent scientists in France, Émile Picard, claimed on that occasion that it was too early to decide for or against the theory.[3] He expressed mixed feelings about the conceptions of space and time, on which the special and the general theories are based, referring to them as metaphysics

3 Biezunski, Michel. "Einstein's Reception in Paris," in *The Comparative Reception of Relativity*, ed. T. Glick (Dordrecht: D. Reidel, 1987), 172.

rather than physics. At the same time, there were also some attempts to refute Einstein's theory. The debate was not strictly scientific or limited to learned circles but involved a part of the general public as well. The attitudes toward Einstein's theory were influenced by prevailing images of science, by scientific ideologies, and by political attitudes toward Germany and German science, only four years after a devastating war.

The general attitude toward Einstein and his theory changed dramatically during his visit to Paris. After a public lecture at the Collège de France, the press almost unanimously admired how clearly he developed his ideas and gave the audience a general feeling of understanding. After the visit, Solovine wrote to Einstein: "Your efforts in Paris were, in fact, extraordinary. But if one takes into consideration the great result you achieved, you will admit that it was worth the effort to come here. The standing of your theories here is entirely different now from before; and regarding personal impressions, people consider themselves very lucky to have made your (personal) acquaintance."[4]

Borel's introduction to the booklet reflects the status of Einstein's ideas and the attitudes toward his theory in the French scientific arena in those days. As was the case in Germany, the theory met with some public opposition because of its revolutionary character, challenging common-sense conceptions of space and time. Another major figure in the French discussion was the mathematician, physicist, and philosopher Henri Poincaré, according to whom a system of scientific formulas has, in principle, an infinite num-

4 Maurice Solovine to Einstein, April 27, 1922, CPAE vol. 13, Doc. 168, p. 153.

ber of different interpretations. But how could one then be sure that Einstein's break with the classical notions of space and time was actually necessary?

Borel's preface is a reaction to these discussions. He writes:

> The translator and the publisher of this book have asked me to present it to the public of French language. It does not need that in any way; because the curiosity provoked including among the large public by the strangeness of the theory of relativity itself and by the violent attacks of which is has occasionally been the object has rendered the name of Einstein so universally famous that one does not have to fear that a work signed by him does not find readers. These readers, once found, will be rapidly seduced and subjugated by the subtlety, the elegance and the force of a thinking always rather sure of itself so that it is not afraid to lower itself to become sometimes informal and to be appealing interchangeably to arguments of common sense and high mathematics.

In what follows, Borel addresses some of the principal objections that came up in the French discussion. His aim is to give a sober assessment of Einstein's achievements, countering the exaggerations that had been associated with it. He stresses, for instance, that the practical value of the theory is very limited: "There is approximately the same numerical relation between the theory of relativity and the ordinary mechanics as there is between the sphericity of the earth and the art of architecture." He also shows himself to be skeptical with regard to the cosmological implications claimed by general relativity, comparing them to an attempt by microscopic beings in a droplet of water to infer from their observations anything about the terrestrial globe and what happens on its surface. He writes: "One has to effectively recognize that Mr. Einstein is far from being the only intellectual who has given in to the temptation to let himself go to such speculations." The objection that there might be another interpretation of Einstein's formulas is countered by the observation that nobody had yet found one and that Einstein's own interpretation therefore emerges, quite in Poincaré's sense, as perhaps not the only possible one but by far the most convenient.

THE ITALIAN TRANSLATION

The mathematician Tullio Levi-Civita recommended to Einstein that he grant permission to the engineer Giuseppe Luigi Calisse to translate the booklet into Italian. Einstein agreed and was very pleased that Levi-Civita promised to write an introduction to this edition.

Levi-Civita was one of the most prominent Italian mathematicians and published a great number of works on pure and applied mathematics. Between 1899 and 1900, Levi-Civita and his tutor Gregorio Ricci-Curbastro wrote a fundamental treatise on absolute differential calculus and its applications to express geometrical and physical laws in Euclidean and non-Euclidean spaces. Einstein and the mathematician Marcel Grossmann later used and developed these new mathematical tools to formulate the general theory of relativity. In the years 1915–1917, Einstein and Levi-Civita corresponded about mathematical problems of the theory.

During the early 1920s, Levi-Civita actively participated in the scientific, ideological, and public debates around the theories of relativity taking place then in Italy. One of the main issues in the debate among mathematicians,

ALBERTO EINSTEIN

SULLA TEORIA SPECIALE E GENERALE DELLA RELATIVITÀ

(VOLGARIZZAZIONE)

Traduzione dal tedesco di G. L. Calisse

Prefazione del Prof. T. Levi-Civita

BOLOGNA
NICOLA ZANICHELLI
EDITORE

mathematical physicists, and astronomers (at that time, there was no chair in theoretical physics at any Italian university) was the revolutionary nature of Einstein's new theory. To many, the view that science might undergo a revolution was unacceptable.

In March 1921, Levi-Civita presented a paper titled "How a Conservative Could Reach a Threshold of the New Mechanics." This paper became very influential and was also translated into French and Spanish. It attempted to show that Einstein's theory of gravitation could be deduced by simple mathematics from the classical formulas of mechanics. Before demonstrating this point, Levi-Civita remarked that no scientist should be fearful of the new, but that researchers had to be conservative; they had to protect established paradigms and be critical of any effort to abolish a successful theory.

A colleague of Levi-Civita's, the mathematical physicist Roberto Marcolongo, who became a strong supporter of relativity, in reaction to the appeal of certain astronomers to save Newton's law, wrote: "The great law of Newton is not in any danger. . . . On the contrary, one of the most beautiful characteristics of the new theories is that they conserve the glorious edifice constructed by Newton, while improving it with modifications, which are qualitatively very slight and conceptually grandiose."[5]

In October 1921, Einstein presented several lectures on the theory of relativity at the universities of Bologna and Padova. There he emphasized the benefits of the evolutionary view of scientific progress. He asserted that the

5 Cited in Reeves, Barbara J. "Einstein Politicized: The Early Reception of Relativity in Italy," in *The Comparative Reception of Relativity*, 197.

theory of relativity was not a revolution but a slow and constrained evolution. In the booklet he wrote (p. 91): "No fairer destiny could be allotted to any physical theory, than that is should of itself point out the way to the introduction of a more comprehensive theory, in which it lives on as a limiting case."

Levi-Civita's introduction to Einstein's booklet was written in this spirit. The emergence of the theory of relativity in 1905 was a result of an obvious extension of known laws and criteria:

> The mathematical scheme of natural philosophy has recently undergone a profound transformation due to the work of Einstein. This transformation which he conceived and which is now universally known as the theory of relativity, was inaugurated in 1905 by uniting an experimental fact with a criterium of relativity. The criterium presents itself under the modest appearance of an obvious extension to the propagation of light of a well-known equivalence, with respect to the laws of mechanics, of two reference frames in uniform motion with respect to each other; but it becomes by this union so loaden with consequences and so fruitful that its recognition justly marks the beginning of a new era in the history of science.

Likewise he claimed that the theory of general relativity was a result of the appropriate mathematical formulation of an extension of the relativity principle and not the result of a replacement of an old principle with a new one: "The successive development of the Einsteinian doctrines (the so-called general relativity) reaches its maturity in 1915 thanks to the systematic employment of the methods of absolute differential calculus by our Ricci. It is still dominated by an extension of the principle of relativity."

Levi-Civita concludes the introduction by saying:

> The speculative importance of relativity is so enormous that in a few years more than 700 works, books, smaller publications and articles, have been dedicated to it. Among them are also writings of great value that have to a large measure contributed to disseminate the new word. But undeniably the public strives to enter into spiritual union with the discoverer. This was rightly perceived by the engineer Calisse and he saw to satisfy the natural desire of our compatriots by a translation of the mentioned volume by Einstein which reflects his thinking in a distinguished and faithful form.

THE SPANISH TRANSLATION

In Spain, as in Italy and France, mathematicians took the main interest in Einstein's theory of relativity, and they participated in the main debates around its reception in the scientific community. The leading mathematician was Rey Pastor, who created a young and vibrant community of scholars engaged in contemporary mathematical research. All the members of this group studied abroad, mostly in Italy. In 1920, Pastor spent a few months in Germany and during that time corresponded with Einstein and invited him to give a series of lectures in Spain. Einstein welcomed the invitation but promised to visit Spain only at a later date, which he did in February 1923, on his way back from Japan and Palestine to Berlin.

In a letter from April 1920, Rey Pastor asked for permission to translate the booklet on the special and general relativity into Spanish. He wrote: "The Sociedad Matemática Española has long had the intention of translating your fine popular book [On] the Special and the Gen. Theory of

LA TEORÍA DE LA RELATIVIDAD

ESPECIAL Y GENERAL

AL ALCANCE DE TODOS

POR

A. EINSTEIN

TRADUCIDA DE LA 14.ª EDICIÓN ALEMANA POR

F. LORENTE DE NO

MADRID
1923

Relativity into the Spanish language in order to acquaint its members with it, and specifically, publish it as a supplement to the society's journal."[6] He repeated this request in August, and Einstein informed him that he had given instructions to his publisher to grant the necessary rights.

The booklet was translated by Lorente de Nó, a student of Pastor's who worked in Rome with Levi-Civita on relativity. Part of it was published serially, with a few pages appearing in each of the subsequent issues of the Spanish-Argentinian journal *Revista Matemática Hispanoamericana* between September 1921 and April 1922. The translation was also published as a book in 1921 in Madrid and again in 1923. Both editions contain only Einstein's text without any introductions or accompanying remarks.

Apart from mathematicians, other communities interested in Einstein and his theory were engineers, physicists, astronomers, and Spanish intellectuals. Such a broad and diverse interest generated a need for professional and popularized presentations of the theory at different levels. Einstein's booklet was only one of several popular texts on relativity, translated from German or written by Spanish scholars. Among other volumes translated from German were books on relativity by the astronomer Erwin Finlay-Freundlich (1885–1964) and the philosopher Friedrich Albert Moritz Schlick. Einstein's booklet was also included on the list of recommended reading for a course on relativity that formed part of the curriculum in schools of engineering.

The Jesuit mathematician Enrique de Rafael developed and taught courses on relativity for engineers. He also wrote reviews on books on relativity, which he published in the popular science journal *Iberica*, issued by the

6 Julio Rey Pastor to Einstein, April 22, 1920, CPAE vol. 9, Doc. 391, p. 328.

Jesuits' Ebro Observatory. Among the reviews was one of Einstein's special and general theory of relativity. Actually, de Rafael reviewed the French edition of the booklet with Emile Borel's introduction (mentioned in the context of the discussion of the French translation). He praised the introduction for its philosophical rigor and for distinguishing between the true and confirmed aspects of Einstein's theory and the nonconfirmed and obscure aspects so that "nobody is detained from acknowledging and admiring the former without being confused by the latter."[7]

Several points in this review deserve attention. Discussing the transition from special to general relativity, de Rafael begins with the equivalence of all reference frames with respect to the laws of nature and refers to it as the bold and ingenious idea that guided Einstein's discovery. Because the review was published in a journal read by astronomers, de Rafael emphasized the theory's success in explaining the shift of the perihelion of Mercury and—referring to the gravitational redshift—quoted the last sentences of Einstein in appendix 3 on the empirical tests of general relativity. Einstein stated there that if this prediction could not be confirmed, then the general theory of relativity would be untenable. However, if its validity could be established, astronomers would have an important source of information on stellar masses.

THE RUSSIAN TRANSLATION

The Russian version of the booklet was published in 1921 by the publishing house Slowo (Word), which operated in Germany. The intended audience was the Russians living

7 The review by Enrique de Rafael of Borel's introduction to the French edition of the booklet was published in *Iberica* 15 (1921): 288.

АЛЬБЕРТЪ ЭЙНШТЕЙНЪ

ТЕОРІЯ ОТНОСИТЕЛЬНОСТИ

ОБЩЕДОСТУПНОЕ ИЗЛОЖЕНІЕ

Переводъ съ нѣмецкаго
Г. Б. Ительсона

1921

КНИГОИЗДАТЕЛЬСТВО «СЛОВО»

outside of Russia rather than those living within the country. At the time, around 400,000 Russians lived in Berlin, and two Russian daily newspapers were printed in Germany. Einstein was very pleased to grant the translation rights to Gregorius Itelson, a Russian logician-philosopher living in Berlin. Itelson was well known in the intellectual circles in Berlin, and Einstein held him in great esteem and affection and authorized him to translate a number of his other works as well. Einstein had spoken favorably of him in a letter to Betty Neumann, his paramour at the time. Responding to Itelson's request, Einstein wrote a special introductory note to the Russian edition. This was one of the very few times when he agreed to do this and the only one in which he included personal remarks about the translator. Years later he claimed to have written the brief foreword only to fulfill the heart's desire of the esteemed Itelson.

The introduction reads:

> More than ever it is necessary in our hectic times to nurture those things which can bring people of a different language and nation closer to each other again. From this point of view it is of particular importance to facilitate the exchange of scientific endeavors even under these currently difficult conditions. I am glad that my booklet will now appear in the Russian language, the more so as Herr Itelson, whom I highly esteem, is guaranteed to provide an excellent translation. The author has often been scolded for saying his booklet is "intelligible to all"; and therefore the Russian reader who encounters difficulties in comprehension should not get angry at himself or Herr Itelson. The really guilty one is none other than the author himself.

In 1926, the 74-year-old Itelson was brutally beaten on a main street in Berlin by anti-Semites screaming "beat the Jews to death."[8] He was taken to the hospital and died a few days later. Einstein attended the funeral. Shortly afterward, Einstein took action to fulfill Itelson's wish to send part of his valuable library to Jerusalem and to sell part of it to provide funds for Itelson's foster daughter. Itelson's library was indeed sent to Jerusalem and was subsequently incorporated into the Hebrew University library.

THE CHINESE TRANSLATION

In the early 1920s, Albert Einstein and the theories of relativity attracted broad interest in China. This interest was cultivated by young Chinese physicists, trained in Japan, Europe, and the United States. One of the most prominent members of this group was Xia Yuanli, who received his bachelor's degree from Yale University and continued his studies in Berlin, where he met Einstein and attended his lectures. Xia was one of the first Chinese theoretical physicists to spread the theory of relativity in China. He taught courses on relativity, wrote newspaper articles, and gave public lectures.

Xia translated Einstein's booklet on the special and general theory of relativity into Chinese. The first edition was published in April 1921 in the special "relativity issue" of the magazine *Gaizao* (The Reconstruction), and in 1922 it was printed as a separate volume by Commercial Press. It became the first book on relativity theory in China and was

8 Cited in Freudenthal, Gideon, and Tatiana Karachentsev. "G. Itelson: A Socratic Philosopher," in *Otto Neurath and the Unity of Science*, ed. J. Symons, O. Pombo, and J. M. Torres (New York: Springer, 2011), 114.

通俗叢書

相對論淺釋

愛因斯丹著

夏文環譯

共學社

1922

broadly influential throughout China and Southeast Asia.[9] Several new editions appeared between 1923 and 1933. To one of them Yuanli added a "Brief Biography of Einstein." In addition to an ordinary account of Einstein's personal and scientific life story, Xia recalls: "When I was in Berlin in 1919, I became acquainted with Einstein through [Max] Planck. I attended Einstein's lectures at Berlin University and he always tried tirelessly to dispel my doubts."

After the booklet was published, readers complained that it was difficult to understand. To respond to these complaints and to help their readers, the editors of *Gaizao* suggested that Xia provide a more accessible explanation of Einstein's ideas. This led to Xia's article "Einstein's Relativity and his Biography," which was published in *Gaizao* in April 1922.

Xia's article begins as follows: "Einstein's theory of relativity is the newest, the most advanced, and the most profound theory of today's physics. For people who are just starting to study this theory, there may be too many subtleties and twists. One is always afraid that he may have picked up some specifics but lost the main theme. Therefore, the first thing I am going to do is to summarize the important ideas of the theory of relativity despite the risk of repetition." On the importance of the theory, he remarks enthusiastically: "The theory of relativity, which was created by the German physicist Einstein, is indeed one of the greatest inventions of human beings. Due to the general theory of relativity, the mysterious gravity also has its new interpretation and physics and geometry become inseparable."

9 Hu, Danian. *China and Albert Einstein: The Reception of the Physicist and His Theory in China 1917–1979* (Cambridge, MA: Harvard University Press, 2005), 91.

LA REKONSTRUO 改造

相對論號

第三卷第八號

要目

中華民國拾年八月拾五日

Xia's article may be viewed as a corollary to his translation of the booklet. It contributed significantly to the public understanding of the theory of relativity.

THE JAPANESE TRANSLATION

The first Japanese edition of Einstein's booklet was published by the Iwanami-Shoten publishing house at the beginning of July 1921 under the title *Lectures on the Relativity Theory*. It enjoyed unusual success—the sixth reprint appeared within a month. It was translated by Ayao Kuwaki and Yosirou Ikeda. Einstein had met Kuwaki in 1909 when he was still living in Bern, Switzerland. Einstein was very grateful to Kuwaki for completing this project. He wrote: "I am extremely pleased that you and your colleague have translated my booklet into Japanese. I still remember well your visit to Berne, especially since you were the first Japanese, indeed the first East Asian whose acquaintance I ever made. You astounded me then with your great theoretical knowledge."[10]

Einstein approved Kuwaki's request to translate more of his work into Japanese. This ambitious undertaking was initiated by another publisher, Kaizosha. The result was a four-volume collection of Einstein's papers and a December 1922 special issue of the periodical *Kaizo* (The Reform). This was the *Taisho period* in Japan, often referred to as *Taisho democracy*, during which Japan opened itself to Western culture, Western ideologies, and science. The process had actually begun in the preceding Meiji period but was greatly accelerated during the Taisho period. *Kaizo*

10 Einstein to Ayao Kuwaki, December 28, 1920, CPAE vol. 10, Doc. 246, p. 342.

理學博士　長岡半太郎序
理學博士　桑木彧雄
理學士　池田芳郎　共譯

アインスタイン 相對性原理講話

東京　岩波書店發行

widely supported the Taisho democracy and was read by Japanese intellectuals. Its editor extended to Einstein an invitation to visit Japan, which Einstein gladly accepted. His 43-day-long visit that began in November 1922 was described by the German Ambassador to Japan as "the parade of a general returning from a triumphant campaign. . . . . the whole Japanese populace from the highest dignitaries down to rickshaw coolies participated spontaneously."[11]

At the time of Einstein's visit to Japan, a social class was already in place that demanded scientific knowledge and showed great interest in Einstein's ideas. In Einstein's letter of invitation, the representative for *Kaizo* wrote that almost every month his journal published articles on the theory of relativity; therefore, "now the interpretation of, or the discussion on the Theory of Relativity is the center of the academic study and interest and it has even the greatest popularity in our country."[12] Einstein received a similar message from Hantaro Nagaoka, whom he had met in Berlin, where the latter had studied mathematical physics with Max Planck. Nagaoka returned to the University of Tokyo and became the doyen of the physical society in Japan. He wrote to Einstein: "Thanks to translations of various writings, popular lectures and books, the Japanese have regarded the principle of relativity with great interest."[13]

Nagaoka wrote a foreword to the Japanese edition of Einstein's booklet, which he begins by describing the general popularity of Einstein and his ideas. He then recalls

11 Cited in Kaneko, Tsutomu. "Einstein and Japanese Intellectuals," in *The Comparative Reception of Relativity*, 352.

12 Kôshin Murobuse to Einstein, before September 27, 1921, vol. 12 (German edition), Doc. 245, p. 289.

13 Hantaro Nagaoka to Einstein, March 26, 1922, CPAE vol. 13, Doc. 115, p. 119.

his own recent visit in England, where a bookseller told him that everyone was reading this booklet. Nagaoka is particularly impressed by the fact that the interest in a theory that was developed in Germany was not affected by the circumstance that England and Germany, just two years earlier, had been fighting a war against each other. He refers to the British astronomers who observed the solar eclipse, confirming Einstein's theory.

Nagaoka also reports about his visit to the United States, where he saw the experimental setup of the Michelson-Morley experiment that led FitzGerald and Lorentz to the length-contraction hypothesis (see section 16 of the booklet). He emphasizes the role of Lorentz in paving the road to special relativity, although his theory "was not as logically or philosophically complete as Einstein's." Nagaoka refers only briefly to the great progress implied by the relativistic theory of gravity.

He quotes from the classical Chinese philosophy book *Wei-shu*, which had been banned for 2000 years by mainstream Confucianism: "The earth moves all the time, but people do not know it. People sit in a closed room of a large ship and they do not feel the ship sailing." It continues: "When we think about this saying, we cannot fail to see the equivalence to Einstein's hypothesis. It is a pity that there have been continuous bans and oppressions to stop the development of various outbursts of nonconventional philosophical and physical ideas."

In conclusion, Nagaoka remarks that although the booklet had been written for ordinary readers, it was still not easy to understand. He hoped that, despite this difficulty, there would be additional publications of translated works of Einstein in Japan and that many Japanese people would eventually grasp the meaning of the relativity principle.

THE POLISH TRANSLATION

As in other countries, the observations during the 1919 solar eclipse confirming the prediction of the gravitational bending of light led to a sudden surge of interest in the theory of relativity among Polish intellectuals. This was reflected in newspaper articles, popular books and brochures, as well as in public debates and disputes. Such activities were particularly numerous in Lwów, a city in which academic and intellectual life flourished. Among the strong adherents of relativity was Maksymilian Tytus Huber, a professor of technical mechanics. A Polish philosopher published an aggressive criticism of Einstein's theory in the popular press, to which Huber responded with a series of five articles published in the same newspaper explaining and defending Einstein and his theory. Other attacks on relativity prompted Huber to deliver a series of popular lectures, but his most important contribution to the public understanding of Einstein's theory was his translation of the booklet, which appeared in November 1921.

Huber wrote a relatively long introduction in which he expressed the hope that his "modest" work of spreading the ideas of a great thinker who paved new roads in science

> will serve for many readers as a necessary supplement to the present brochure, which is not "popular" in the usual sense, but rather, as someone said jokingly, "popular with physicists." This booklet cannot serve as after dinner reading, even for a scientifically educated mind, but it may provide a reader who does not spare intellectual effort, moments of deep spiritual satisfaction, which are sensed by a researcher who succeeds in understanding a great secret of Nature. The path leading to the summits of the theory of relativity along which its creator

A. EINSTEIN.

O SZCZEGÓLNEJ I OGÓLNEJ
TEORJI WZGLĘDNOŚCI

(WYKŁAD PRZYSTĘPNY)

Z UPOWAŻNIENIEM AUTORA
PRZEŁOŻYŁ Z 11-GO WYDANIA ORYGINAŁU
INŻ. DR. M. T. HUBER
PROFESOR POLITECHNIKI LWOWSKIEJ

WYDANIE DRUGIE
PRZEJRZANE I UZUPEŁNIONE DJALOGIEM O ZARZUTACH
PRZECIWKO TEORJI.

LWÓW – WARSZAWA.
KSIĄŻNICA POLSKA TOWARZYSTWA NAUCZYCIELI SZKÓŁ WYŻSZYCH.
MCMXXII.

is leading us is not easy, but who if not he himself can better illuminate this uniform image of the world, which reveals itself from those summits to the eyes of our soul.

Huber's introduction includes a biography of Einstein, but the main emphasis is on the different kinds of objections to the theory of relativity:

In parallel with the interest in intelligent circles, there are more and more complaints about the difficulty and inaccessibility of the theory. In fact, those who do not understand it . . . cross the lines to the camp of the opponents, hiding under the wings of some political system or joining some (though few) physicists whose attitude to the theory of relativity is still skeptical. One often hears from such individuals the well-known, but unfounded, general statement that great scientific discoveries are marked by their simplicity and therefore by their accessibility. More than half a century has passed since the establishment of Maxwell's electromagnetic theory of light, one of the most prominent achievements of the previous century, and up until today this theory has not become part of the school curriculum. Is it possible to explain it clearly and comprehensibly to the nonmathematical mind? The history of knowledge teaches us that historical achievements of human thought were generally met with the opposition of contemporaries. This is also the case of the theory of relativity in which psychological elements also play a role.

Huber also emphasizes the anti-Semitic element of part of the opposition to Einstein and his ideas, both in Germany and in Poland. On this he writes:

> His conciliatory position in the name of the international nature of pure knowledge and partly also hostility of some scientists *minorum gentium*, who simply did not understand the theory of relativity, caused the outbreak of hostile demonstrations against him. This can be understood against the background of the war and—in the sense of a French proverb—even the Prussian nationalists can be forgiven who unload their anger at Germany's defeat through the paroxysm of anti-Semitism. But it is impossible to remain indifferent to such symptoms here, which present real danger to our national culture. What could a judgment of scientific achievements through the prism of racial or national prejudices lead to? I observed such judgment

recently on the occasion of the initial general interest in the theory of relativity in Lwów.

Huber concludes his introduction with the following wish: "May the present publication be a good beginning in breaking the ice of naïve prejudices and pseudo-philosophical superstition for the benefit of our scientific culture in the reborn and reunited Fatherland." He refers here to the reunification of Poland after more than 120 years of partition among Austria, Germany, and Russia.

A second Polish edition appeared a year later, in 1922. In the second edition, Huber added, as an appendix, Einstein's article from 1918, written as a dialogue between a "relativist" who responds to questions and the critical remarks of a "critic"[14]

THE CZECH TRANSLATION

A Czech translation of Einstein's book was published in 1923. This was another one of the few cases in which Einstein wrote a special foreword. He did it with pleasure, remembering the time he spent as a professor in the German department of the Charles University in Prague between April 1909 and August 1912.

Einstein's work in Prague was an important step toward his completion of the theory of general relativity. There he wrote 11 scientific papers, 6 of which were devoted to relativity. In the first of these papers, in 1911, he discussed the bending of light and the gravitational redshift, which he had already deduced from the equivalence principle in

14 Huber, Maksymilian Tytus. "Dialogue about Objections to the Theory of Relativity." *Die Naturwissenschaften* 6 (1918). English translation in CPAE vol. 7, Doc. 13.

ALBERT
EINSTEIN

Theorie
RELATIVITY
SPECIELNÍ i OBECNÁ

Nakladatel
FR. BOROVÝ v PRAZE

1907, and now explored them as observable effects. When he returned to Zurich, he realized that the gravitational field is represented by the geometrical properties of curved space-time. This was a landmark in the development of general relativity.

In his foreword to the Czech edition, Einstein described his work in Prague:

> I am glad that this little booklet, in which the basic ideas of the theory of relativity are presented without mathematical formalism, appears now in the national language of the country where I found the necessary contemplation to gradually give the general theory of relativity a more precise form, an endeavor whose basic idea I adopted already in 1908 (he must have meant 1907). In the quiet rooms of the Institute of Theoretical Physics at the German University of Prague, in Vinicna ulice, I discovered in 1911 that the principle of equivalence demands a deflection of the light rays passing by the sun with observable magnitude—this without knowing that more than one hundred years ago a similar consequence had been anticipated from Newton's mechanics in combination with Newton's emission theory of light. I also discovered at Prague that still not completely confirmed consequence of the redshift of spectral lines. However, only after my return in 1912 to Zurich did I hit upon the decisive idea about the analogy between the mathematical problem connected with my theory and the theory of surfaces by Gauss—originally without knowledge of the research by Riemann, Ricci, and Levi-Civita. The latter research came to my attention only through my friend Grossmann in Zurich when I posed the problem to him only to find generally covariant tensors whose components depend only upon the derivatives of the coefficients of the quadratic

fundamental invariant. Today it appears that we can clearly recognize the achievements and limitations of the theory. The theory provides deep insights into the physical nature of space, time, matter and gravitation, but no adequate means to solve the problems of quanta and the atomic constitution of elementary electric structures that constitute matter.

THE HEBREW TRANSLATION

The idea to translate Einstein's book into Hebrew had already surfaced in 1921 when the "interterritorial" publishers Renaissance wrote to Einstein's secretary—at that time his stepdaughter Ilse—inquiring about the translation rights into Hebrew and Yiddish. Ilse forwarded this request to Friedrich Vieweg, director of the Vieweg publishing company. She added: "It would be a special favor to Prof. Einstein if you could agree not to demand any compensation for the translation into these two languages, which Prof. Einstein, for his part, intends to forgo."[15] We do not know the response of the publisher, but this proposal did not materialize.

In February 1926 Einstein received a request from Jakob Gruenberg of Vienna for permission to translate Einstein's booklet into Hebrew.[16] Gruenberg introduced himself as a holder of a doctoral degree from the University of Heidelberg, where he had studied mathematics, physics,

15 The Secretary (Ilse Einstein) to Friedrich Vieweg, October 5, 1921, in CPAE vol. 12, Doc. 258, p. 166.

16 Unpublished letter from Jakob Gruenberg to Einstein, July 2, 1926.

על תורת היחסיות הפרטית והכללית

(הרצאה פופולרית)

מאת

אלברט אינשטין

תרגם ברשיון המחבר

יעקב גרינברג

הוצאת «דביר» תל־אביב

תרפ״ט

and chemistry. He had already written about Einstein and his theory for the *Enzyklopädie des Judentums.* This was one year after the opening of the Hebrew University in Jerusalem, of which Einstein was one of the founders, a member of its international board of governors, and head of the academic committee of the board. It is no wonder then that Einstein wholeheartedly welcomed this proposal. Apparently, in this case he took care of the matter himself without involving the Vieweg publishing house. In a follow-up letter in July, Gruenberg informed Einstein that a publisher in Tel Aviv was willing to undertake the project and asked him, also on behalf of the publisher, for an introduction to the Hebrew edition.

On the back of this letter, Gruenberg proposed the wording for the introduction, which Einstein adopted without adding any changes. It reads: "The publication of this book of mine in the language of our forefathers fills my heart with special joy. It is a sign of the change that occurred in this language. It is not confined any more to expressing matters related to our people for our people, but is ready to encompass everything that is of interest for mankind. It serves as an important factor in our strive for cultural independence."

In his own introduction, Gruenberg expresses his excitement with the fact that the Hebrew language will now serve as an arena for the ideas of Einstein, who cherishes this language not only as the language of his forefathers but also as the language of cultural renaissance. He describes the difficulties and the challenges of translating such a sophisticated scientific text into a language that had not been spoken for thousands of years and which was then in the initial stages of its adaptation to scientific discourse.

CONCLUDING REMARKS

This short overview of some of the translations of Einstein's booklet thus confirms what more extensive studies of the comparative reception of relativity have taught us: the reception of a new scientific theory is not a passive absorption of information but an active appropriation and often an intellectual struggle. This appropriation is shaped by the prior shared knowledge of a scientific and public community, by its social structures—for instance, the degree of specialization—and by the status and value ascribed to science itself in a given society. A new scientific theory thus unavoidably enters a field of tension in which it may become the ally or foe of some preexisting stakeholders.

This was particularly the case for Einstein's theories of relativity. Their worldwide spread in a time of both fervent nationalism and internationalism, just after the Great War, was at the same time a litmus test for the role of science in global modernization processes. Was science just a useful tool for technical progress in a fierce competition among nations? Or was it a joint enterprise of humanity based on cooperation that was aimed at enriching shared knowledge? When we browse the impressive collection of foreign translations of Einstein's booklet, we feel that such general questions must have moved its editors and readers.

Einstein's special and general theories of relativity challenged the public understanding of science in other ways as well. They did not square with the widespread image of science whereby progress is gauged by ever more refined measurements or calculations or by building floor by floor an ever more imposing ivory tower that nevertheless stands on the firm ground of common knowledge. Instead, Einstein's theories of relativity challenged precisely this common ground, and their challenges were not so much premised by precision measurements as they were, against all odds, confirmed by novel observations and experiments. While to some contemporary readers this just meant that Einstein's theories were incomprehensible or simply absurd, others could—and can—learn from this booklet the extent to which science is also a matter of questioning deeply ingrained prejudices and of independent thinking.

Appended Documents

A LETTER FROM WALTHER RATHENAU TO EINSTEIN

Einstein sent the booklet to his friend Walther Rathenau, a Jewish-German industrialist, writer, and statesman who, after World War I, served as the minister of foreign affairs of the Weimar Republic. On July 24, 1922, he was assassinated by right-wing elements in Germany.

Rathenau admits in his letter that he has read only the first part of the booklet (the part on special relativity). The letter reflects his wit and sense of humor together with his friendly and appreciative attitude toward Einstein.

From Walther Rathenau

[Berlin,] 10–11 May 1917

Dear and esteemed Mr. Einstein,

I have been immersed in your ideas for weeks; I had barely finished the evangelist Schlick when the *verba magistri* arrived, which are now before me and for which I thank you wholeheartedly.

First a preliminary remark, which is not meant to be a platitude: the prophet is clearer than the evangelist. I would not have thought it possible to force such a radical rearrangement of ideas through, the way you do, with such simple means and using such classical architectonics—I underscore the word classical, in contrast to your "bumpy."

I have read to p. 39 and do not say that it comes easily to me, but certainly relatively easily—as everything that you touch becomes relative. Perhaps I am complicating the matter for myself because all sorts of rudimentary ideas from various sources have led me within the proximity of your force field, and because now I must take in the radiating effects and assimilate them within the existing chains of reasoning.

Shall I tell you a bit about such rudimentary thoughts? Within the light of your halo they will appear as pitiful unmasked ghosts—but maybe I can extend my bizarre thanks to you for my joy and admiration by making you smile for a minute or two. Will the things still occur to me now, around midnight? Let us enumerate, then it will work.

1. Gyroscopes always seemed senseless to me. When built with precision, how does it know that it is rotating? How does it distinguish the direction in space in which it does not want to let itself be tilted? Even if I put it in a box and make it blind, it knows where the polestar is. I have always had the secret feeling that it rotates only when it has a spectator. But if so—it would then have to protect itself with counterforces against the approach of such spectators from infinity. Are there such forces?

2. Ever since the arrival of vestibule trains, I have found walking in the corridors not only an ordeal but also a problem, hence a pleasure. Often I imagined a vestibule train that extended from Berlin not quite all the way to Paris but only to St. Quentin, and within it a smaller one to Verviers, etc. Then one would arrive quite quickly in Paris. Well, there is an end to this anyway—so I do not have much to lose.

3. The smaller insects are, the faster they move. It is customary to feel sorry for dayflies. I told myself sometimes: Maybe it is not so bad; in the end, time diminishes with mass.—Or is it only the sense of time?—If one had to play the

allegro to the 5th [symphony] for such a gnat, it would be over in one minute, otherwise it would mistake it for a funeral march in C major.

Well, time really does depend upon motion! But we do not notice it! And motorcar driving was a fleeting pleasure nonetheless.

(Now it is getting more and more nonsensical and muddled; I think you should not read on. But ultimately, all thoughts arise from elements of lunacy; it is only the critical cement which is lacking here to glue them together.)

4. One thing which is entirely unrelated and yet which seems to me to have a remote connection: I have always had an emotional resistance to the entropy business. It seems to me as if it were correct only inside a bathtub. And moreover: What happens to rays of light that go out into the distance without ever meeting an object? As long as there is a medium: fine. But why should it not come to an end—or be interrupted?

5. A metaphysical frivolity: My senses tell me that everything is at rest in the absolute. Illustration—somewhat in our Lord's spirit: a trip to Italy. I travel the length and breadth of the land, experience it minute by minute, register it, and cut it up into sections. I come home and have a concept of: Italy. An impression like the taste of a fruit or the character of a woman. I can scroll through it again (using the memory sections) and do so when I want to answer individual questions. But otherwise, Italy *rests* within me, is present, alive, and yet motionless. I possess the whole (unfortunately only figuratively speaking, since travel and life are finite)—or I possess at least *a* whole.

Your illustration of the two flashes of lightning and the train really gripped me here (incidentally, I turn it into two dynamite explosions and a czar train). What startles the czar twice is only *a single* matter for the assassin. His rest (in both senses) is greater. Now I expand further. The assassin stands outside the train. Now I place him beyond the Earth's rotation; then beyond the Earth's orbit; then beyond the translation; then beyond—etc. Will the man not be surrounded by an ever deeper silence?

Another sidetrack: Time (epistemologically) dissipates. It only comes into existence, so to speak, through motion. Motion, however, has time as its precondition again, for it is $\frac{s}{t}$. Would we not have to find a way out (epistemologically) by conceiving of motion as a manifestation of force (thus obversely to normally), so we merely note that elements containing varying amounts of force (metaphysical charge) exist—thus ultimately a kind of monadology would emerge?

Now, enough of this. Instead of cheerful thanks I have given you a *pathologia mentalis*, which must have appalled you. Nevertheless, I am sending the letter off, because it has to prove one thing to you *in corpore viti*: the forceful effect of your ideas on a poorly shielded brain. A steel helmet [*Stahlhelm*] is necessary, or at least a sun hat, in order to hold the balance.

My regards in amicable admiration, yours,

Rathenau.

A SAMPLE PAGE OF EINSTEIN'S HANDWRITING

The page in Einstein's handwriting, reproduced here by courtesy of the Albert Einstein Archives at the Hebrew University of Jerusalem, is the only page of the original manuscript to survive. It contains part of section 31, beginning with the text following the equation on page 124.

(3)

finden, d. h. einen Wert der kleiner ist als π, und zwar umso erheblicher, je grösser der Radius des Kreises im Vergleich zum Radius R der „Kugelwelt" ist. Aus dieser Beziehung können die Kugelgeschöpfe den Radius R ihrer Welt bestimmen, auch wenn ihnen nur ein relativ geringer Teil ihrer Kugelwelt für ihre Messungen zur Verfügung steht. Ist aber dieser Teil allzu klein, so können sie nicht mehr konstatieren, dass sie sich auf einer Kugelwelt und nicht auf einer euklidischen Ebene befinden; ein kleines Stück einer Kugelfläche unterscheidet sich wenig von einem gleich grossen Stück einer Ebene.

Wenn also die Kugelgeschöpfe auf einem Planeten wohnen, dessen Sonnensystem nur einen verschwindend kleinen Teil der Kugelwelt einnimmt, so haben sie keine Möglichkeit, darüber zu entscheiden, ob sie in einer endlichen Welt oder einer unendlichen Welt leben, weil das Stück Welt, was ihrer Erfahrung zugänglich ist, in beiden Fällen praktisch eben, bezw. euklidisch ist. Die Anschauung zeigt unmittelbar, dass für unsere Kugelgeschöpfe der Kreisumfang mit dem Radius zunächst bis zum „Weltumfange" wächst, um dann bei noch weiter wachsendem Radius allmählich wieder bis zu null abzunehmen. Die Kreisfläche wächst dabei immer mehr, bis sie schliesslich gleich wird der Gesamtfläche der ganzen Kugelwelt.

Vielleicht wird sich der Leser wundern, dass wir unsere Geschöpfe gerade auf eine Kugel und nicht auf eine andere geschlossene Fläche gesetzt haben. Aber dies hat seine Berechtigung, weil die Kugel gegenüber allen anderen geschlossenen Flächen durch die Eigenschaft ausgezeichnet ist, dass all ihre Punkte gleichwertig sind. Das Verhältnis des Umfanges u eines Kreises zu seinem Radius r ist zwar von r abhängig, aber bei gegebenem r für alle Punkte der Kugelwelt das gleiche; die Kugelwelt ist eine „Fläche konstanter Krümmung").

Es gibt zu dieser zweidimensionalen Kugelwelt ein dreidimensionales Analogon, den dreidimensionalen sphärischen Raum, welcher von Riemann entdeckt worden ist. Seine Punkte sind ebenfalls alle gleichwertig. Er besitzt ein endliches Volumen, welches durch seinen „Radius" R bestimmt ist ($2\pi R^3$). Kann man sich einen sphärischen Raum vorstellen? Sich einen Raum vorstellen heisst nichts anderes als sich einen Inbegriff „räumlicher" Erfahrungen vorstellen, d. h. von Erfahrungen, die man beim Bewegen „starrer" Körper haben kann. In diesem Sinn ist ein sphärischer Raum vorstellbar.

Um einen Punkt aus ziehen wir Gerade (spannen wir Schnüre) nach allen Richtungen und tragen auf jeder derselben die Strecke r mit dem Masstabe auf. Alle freien Endpunkte dieser Strecken liegen auf einer Kugelfläche. Die Fläche dieser (F) können wir mit einem Masstabquadrat besonders ausmessen. Ist die Welt euklidisch, so ist $F = \pi r^2$; ist die Welt sphärisch, so ist F stets kleiner als πr^2. F wächst mit wachsendem r bis zu einem durch den „Weltradius" bestimmten Maximum, um bei weiter wachsendem Kugelradius r allmählich wieder bis zu null abzunehmen. Die vom Mittelpunkt ausgehenden radialen Geraden entfernen sich zunächst immer weiter

MANUSCRIPT OF APPENDIX 3 OF THE BOOKLET

Appendix 3 of the booklet, on the experimental confirmation of the general theory of relativity, was written by Einstein for the first English edition at the request of the translator, Robert Lawson. It was also included in the 10th German edition. The manuscript was sent to the translator and changed hands until it was donated to Oglethorpe University in 1934 by Nellie Gaertner in memory of her father, Herman Julius Gaertner, who was instrumental in the reestablishment of the university on its present campus. He was in the first faculty hired in 1915, and his career in education, graduate studies, German, mathematics, and psychology at the university spanned 34 years.

1.)

Über die Bestätigung der allgemeinen Relativitäts-Theorie durch die Erfahrung.

Den Prozess des Werdens einer Erfahrungswissenschaft denkt man sich bei schematisch erkenntnistheoretischer Betrachtungsweise als einen fortgesetzten Induktionsprozess. Die Theorien erscheinen als Zusammenfassungen einer grossen Menge von Einzelerfahrungen in Erfahrungsgesetze, aus denen durch Vergleichung die allgemeinen Gesetze ermittelt werden. Die Entwicklung der Wissenschaft erscheint von diesem Standpunkte aus ähnlich einem Katalogisierungs-Werk, als ein Werk der blossen Empirie.

Diese Auffassung erschöpft aber den wirklichen Prozess keineswegs. Sie verschweigt nämlich die wichtige Rolle, welche Intuition und deduktives Denken in der Entwicklung der exakten Wissenschaft spielen. Sobald nämlich eine Wissenschaft über das primitivste Stadium herausgekommen ist, entstehen die theoretischen Fortschritte nicht mehr durch eine bloss ordnende Thätigkeit. Der Forscher entwickelt vielmehr, angeregt durch Erfahrungs-Thatsachen, ein Gedankensystem, das logisch auf eine meist geringe Zahl von Grundannahmen, die sogenannten Axiome, aufgebaut ist. Ein solches Gedankensystem nennen wir eine Theorie. Die Theorie schöpft ihre Daseinsberechtigung daraus, dass sie eine grössere Zahl von Einzelerfahrungen verknüpft; hierin liegt ihre „Wahrheit".

Es kann nun zu demselben Komplex von Erfahrungsthatsachen verschiedene Theorien geben, die sich sehr bedeutend voneinander unterscheiden. Die Übereinstimmung der Theorien in den der Erfahrung zugänglichen Konsequenzen kann eine so weitgehende sein, dass es schwer fällt, Konsequenzen zu finden, bezüglich welcher sich die beiden Theorien von einander unterscheiden. Ein solcher Fall von allgemeinem Interesse liegt beispielsweise auf dem Gebiete der Biologie vor in der Darwin'schen Theorie der Entwicklung der Arten durch Zuchtwahl im Kampf ums Dasein und in derjenigen Entwicklungstheorie, die sich auf die Hypothese von der Vererbung erworbener Eigenschaften gründet.

Ein derartiger Fall von weitgehender Übereinstimmung der Konsequenzen liegt vor bei der Newton'schen Mechanik einerseits und der allgemeinen Relativitätstheorie andererseits. Diese Übereinstimmung geht so weit, dass bisher nur wenige der Erfahrung zugängliche Folgerungen der allgemeinen Relativitätstheorie haben gefunden werden können, zu denen die frühere Physik nicht führt – trotz der tiefgehenden Verschieden-

Einstein manuscript, dated between February 4 and April 22, 1920, Courtesy of the Archives, Philip Weltner Library, Oglethorpe University (Atlanta, GA)

2)

heit der Grundvoraussetzungen der Theorien. Diese wichtigen Konsequenzen wollen wir hier noch einmal betrachten und auch die bisher darüber gesammelten Erfahrungen kurz besprechen.

1. Die Perihel-Bewegung des Merkur.

Nach der Newton'schen Mechanik und dem Newton'schen Gravitationsgesetz würde ein einziger um eine Sonne kreisender Planet eine Ellipse um die Sonne (bezw. um den gemeinsamen Schwerpunkt von Sonne und Planet) beschreiben. Die Sonne (bezw. der gemeinsame Schwerpunkt) liegt hiebei in dem einen Brennpunkt der Bahnellipse, derart, dass der Abstand Sonne – Planet im Laufe eines Planetenjahres von einem Minimum zu einem Maximum wächst und dann wieder zu dem Minimum zurückgeht. Setzt man statt des Newton'schen Anziehungsgesetzes ein etwas anderes in die Rechnung ein, so findet man, dass die Bewegung nach diesem Gesetze immer noch so stattfinden würde, dass der Abstand Sonne – Planet periodisch hin und her schwankt; aber der bei einer solchen Periode (von Perihel [illegible] zu Perihel) von der Linie Sonne – Planet beschriebene Winkel würde von 360° abweichen. Die Linie der Bahn wäre dann keine geschlossene, sondern würde im Laufe der Zeit einen ringförmigen Teil der Bahnebene (zwischen dem Kreise des kleinsten und dem Kreise des grössten Planetenabstandes) ausfüllen.

Nach der allgemeinen Relativitätstheorie, welche ja von der Newton'schen etwas abweicht, findet eine derartige kleine Abweichung von der Kepler – Newton'schen Bahnbewegung statt, derart, dass der vom Radius Sonne – Planet zwischen einem Perihel und dem folgenden von einem vollen Umlaufswinkel (d. h. vom Winkel 2π in dem in der Physik üblichen absoluten Winkelmasse) um

$$+\frac{24\pi^3 a^2}{T^2c^2(1-e^2)}$$

abweicht. (Hiebei ist a die grosse Halbachse der Ellipse, e deren Exzentrizität, c die Lichtgeschwindigkeit, T die Umlaufsdauer.) Man kann dies auch so ausdrücken: Nach der allgemeinen Relativität rotiert die grosse Achse der Ellipse im Sinne der Bahnbewegung um die Sonne. Diese Drehung soll nach der Theorie beim Planeten Merkur 43 Bogen-Sekunden in 100 Jahren betragen, bei den anderen Planeten unserer Sonne aber so klein sein, dass sie sich der Konstatierung entziehen muss.

Thatsächlich haben die Astronomen gefunden, dass die Newton'sche Theorie nicht ausreicht, um die beobachtete Bewegung des Merkur mit der der heutigen Beobachtung zugänglichen Exaktheit zu berechnen. Nach Berücksichtigung aller störenden Einflüsse, welche die übrigen Planeten auf Merkur ausüben, zeigte es sich (Leverrier 1859 und Newcomb 1895), dass eine unerklärte Perihelbewegung der Merkurbahn

3.)

übrig blieb, welche sich von den oben genannten + 43 Sekunden pro Jahrhundert nicht merklich unterscheidet. Die Unsicherheit der empirischen Resultate beträgt wenige Sekunden.

2. Die Licht-Ablenkung durch das Gravitationsfeld nach der allgemeinen Relativitätstheorie.

Im § 22 ist dargelegt, dass ein Lichtstrahl durch ein Gravitationsfeld eine Krümmung erfahren muss, welche der Krümmung ähnlich ist, welche die Bahn eines durch das Gravitationsfeld geschleuderten Körpers erfahren muss. Ein an einem Himmelskörper vorbei gehender Lichtstrahl wird nach der Theorie nach diesem hin abgebogen; dieser Ablenkungswinkel α soll bei einem Lichtstrahle, der in einem Abstande von Δ Sonnenradien an dieser vorbeigeht

$$\alpha = \frac{1{,}7 \text{ Sekunden}}{\Delta}$$

betragen. Es sei hinzugefügt, dass diese Ablenkung nach der Theorie zur Hälfte durch das (Newton'sche) Anziehungsfeld der Sonne, zur Hälfte durch die von der Sonne bewirkte geometrische Modifikation („Krümmung") des Raumes erzeugt wird.

Dies Ergebnis erlaubt eine experimentelle Prüfung durch photographische Sternaufnahmen während einer totalen Sonnenfinsternis. Letztere muss nur deshalb abgewartet werden, weil zu jeder andern Zeit die durch das Sonnenlicht bestrahlte Atmosphäre so stark leuchtet, dass die sonnennahen Sterne unsichtbar sind. Die zu erwartende Erscheinung ergibt sich leicht aus nebenstehender Figur. Wäre die Sonne S nicht vorhanden, so würde man einen praktisch unendlich weiten Stern in der Richtung R_1 sehen. Infolge der Ablenkung durch die Sonne sieht man ihn aber in der Richtung R_2, das heisst in einer etwas grösseren Entfernung von der Sonnenmitte, als der Wirklichkeit entspricht.

In Praxi geschieht die Prüfung in folgender Weise. Die Sterne in der Umgebung der Sonne werden bei einer Sonnenfinsternis photographiert. Es wird ferner eine zweite photographische Aufnahme derselben Sterne hergestellt, wenn die Sonne an einer andern Stelle des Himmels ist (d. h. einige Monate später oder früher). Die bei der Sonnenfinsternis aufgenommenen Sternbilder müssen dann gegenüber der Vergleichsaufnahme radial nach aussen (vom Sonnenmittelpunkte weg) verschoben sein um einen Betrag, der dem Winkel α entspricht.

Der Astronomical Royal Society verdanken wir die Prüfung dieses wichtigen Ergebnisses. Ohne sich durch den Krieg und durch diesen geschaffene Schwierigkeiten psychologischer Art beirren zu lassen

4)

hat sie mehrere ihrer bedeutendsten Astronomen (Eddington, Crommelin, Davidson) ausgesandt und zwei Expeditionen ausgerüstet, um bei der Sonnenfinsternis vom 29. Mai 1919 in Sobral (Brasilien) und auf der Insel Principe (Westafrika) die photographischen Aufnahmen zu machen. Die zu erwartenden relativen Abweichungen der Sonnenfinsternisaufnahmen gegenüber den Vergleichsaufnahmen bezüglich der Sternörter auf der Vergleichsaufnahme betragen nur wenige hundertstel Millimeter. Die Anforderungen, welche an die Präzision der Aufnahmen und Vermessung gestellt wurden, waren also keine geringen.

Das Ergebnis der Messung bestätigte die Theorie in durchaus befriedigender Weise. Die Komponenten der beobachteten und berechneten Abweichungen der Sterne (in Bogensekunden) sind in folgender Tabelle enthalten:

Nummer des Sterns	1. Koordinate		2. Koordinate	
	Beobachtet	Berechnet	Beobachtet	Berechnet

3) Die Rotverschiebung der Spektrallinien.

In § 23 ist gezeigt, dass in einem gegen ein galileisches System K rotierenden System K' die Geschwindigkeit des Ganges ruhender gleichbeschaffener Uhren vom Orte abhängig ist. Wir wollen diese Abhängigkeit quantitativ untersuchen. Eine Uhr, die im Abstande r vom Zentrum der Scheibe angeordnet ist, hat relativ zu K die Geschwindigkeit

$$v = \omega r,$$

wenn ω die Rotationsgeschwindigkeit der Scheibe (K') gegenüber K bezeichnet. Bezeichnet ν_0 die Zahl der Schläge der Uhr pro Zeiteinheit (Ganggeschwindigkeit) relativ zu K, falls die Uhr unbewegt ist, so ist die Ganggeschwindigkeit ν der relativ zu K mit der Geschwindigkeit v bewegten, relativ zur Scheibe ruhenden Uhr gemäss § 12

$$\nu = \nu_0 \sqrt{1 - \frac{v^2}{c^2}}$$

oder mit hinreichender Genauigkeit

$$\nu = \nu_0 \left(1 - \frac{1}{2} \frac{v^2}{c^2}\right)$$

oder auch gleich

$$\nu = \nu_0 \left(1 - \frac{1}{c^2} \frac{\omega^2 r^2}{2}\right)$$

Bezeichnet man mit $+\Phi$ die Differenz des Potentials der Zentrifugalkraft zwischen dem Standort der Uhr und dem Scheibenmittelpunkt, d. h. die negativ genommene Arbeit, welche man entgegen der Zentrifugalkraft der Masseneinheit zuführen muss, um sie vom Standpunkt der Uhr auf der bewegten Scheibe zum Mittelpunkt zu transportieren, so ist

5)

$$\Phi = -\frac{\omega^2 r^2}{2},$$

sodass man hat

$$\nu = \nu_0 \left(1 + \frac{\Phi}{c^2}\right)$$

Hieraus ersieht man zunächst, dass zwei gleich beschaffene Uhren in verschiedenem Abstand vom Scheibenmittelpunkt verschieden rasch laufen, welches Ergebnis auch vom Standpunkte eines mit der Scheibe rotierenden Beobachters Gültigkeit hat.

Da nun ~~vom Standpunkte~~ – von der Scheibe aus beurteilt – ein Gravitationsfeld existiert, dessen Potential Φ ist, so wird das gewonnene Resultat überhaupt für Gravitationsfelder gelten. Da wir ferner ein Spektrallinien emittierendes Atom als eine Uhr ansehen dürfen, so gilt der Satz:

Ein Atom absorbiert bezw. emittiert eine Frequenz, welche vom Potential des Gravitationsfeldes abhängt, in welchem es sich befindet. ~~Beobachtet man~~

Die Frequenz eines Atoms, das sich an der Oberfläche eines Himmelskörpers befindet, ist etwas kleiner als die Frequenz eines Atomes des gleichen Elementes, das sich im freien Weltraume (oder an der Oberfläche eines kleineren Weltkörpers) befindet. Da $\Phi = -\frac{KM}{r}$ ist, wobei K die Newton'sche Gravitationskonstante, M die Masse des Himmelskörpers, r den Abstand des Atoms vom Mittelpunkte des Himmelskörpers bedeutet, so müsste eine Rotverschiebung der an der Oberfläche von Sternen erzeugten Spektrallinien gegenüber den an der Erdoberfläche erzeugten Spektrallinien im Betrage

$$\frac{\nu - \nu_0}{\nu_0} = -\frac{K}{c^2}\frac{M}{r}$$

stattfinden.

Bei der Sonne beträgt die zu erwartende Rotverschiebung etwa zwei Milliontel der Wellenlänge. Bei den Fixsternen ist eine zuverlässige Berechnung nicht möglich, weil weder die Masse M noch der Radius r im Allgemeinen bekannt sind.

Ob dieser Effekt thatsächlich existiert, ist eine offene Frage, an deren Beantwortung gegenwärtig von den Astronomen mit grossem Eifer gearbeitet wird. Bei der Sonne ist die Existenz des Effektes wegen seiner Kleinheit schwer zu beurteilen. Während Grebe und Bachem (Bonn) auf Grund ihrer ~~[illegible]~~ eigenen Messungen sowie derjenigen von Evershed und Schwarzschild an der Cyanbande die Existenz des Effektes für sicher gestellt halten, sind andere Forscher insbesondere St. John auf Grund ihrer Messungen der entgegengesetzten Ansicht.

Bei den statistischen Untersuchungen an den Fixsternen sind mittlere Linienverschiebungen nach der langwelligen Spektralseite sicher vorhanden. Aber die bisherige Bearbeitung des Materials erlaubt noch keine sichere Entscheidung darüber, ob jene Verschiebungen

(6.)

wirklich auf die Wirkung der Gravitation zurückzuführen sind. Eine Zusammenstellung des Beobachtungsmaterials nebst eingehender Diskussion vom Standpunkt der uns hier interessierenden Frage findet man in der Abhandlung von E. Freundlich „Zur Prüfung der allgemeinen Relativitätstheorie" (Die Naturwissenschaften 1919 Heft 35 S. 629 Verl. Julius Springer, Berlin).

Jedenfalls werden die nächsten Jahre die sichere Entscheidung bringen. Wenn die Rotverschiebung der Spektrallinien durch das Gravitationspotential nicht existierte, wäre die allgemeine Relativitätstheorie unhaltbar. Andererseits wird das Studium der Linienverschiebung, wenn ihr Ursprung aus dem Gravitationspotential sicher gestellt sein wird, wichtige Aufschlüsse über die Masse der Himmelskörper liefern.

A. Einstein.

FURTHER READING

Other publications by Einstein with which he sought to give comprehensible introductions to his work include the following:

Einstein, Albert. 1922. *The Meaning of Relativity: Four Lectures Delivered at Princeton University, May 1921 by Albert Einstein*. Translated by Edwin Plimpton Adams (1st ed.). London: Methuen.

———. 1992. *Autobiographical Notes. A Centennial Edition*. Edited by Paul A. Schilpp. La Salle, IL: Open Court.

———. 1950. *Out of My Later Years*. New York: Philosophical Library.

———. 1954. *Ideas and Opinions: Based on "Mein Weltbild."* Edited by Carl Seelig and other sources. New York: Bonanza Books.

Einstein, Albert, and Leopold Infeld. 1938. *The Evolution of Physics: The Growth of Ideas from Early Concepts to Relativity and Quanta*. New York: Simon & Schuster.

For accessible introductions to special and general relativity, see the following:

Giulini, Domenico. 2005. *Special Relativity: A First Encounter; 100 Years since Einstein*. New Haven, CT: Oxford University Press.

Schutz, Bernard. 2004. *Gravity from the Ground Up*. Cambridge: Cambridge University Press.

For works on the history of general relativity, see the following:

Howard, Don, and John Stachel (series eds.). 1989–. *Einstein Studies.* Center for Einstein Studies, Boston. New York: Birkhäuser

Renn, Jürgen (ed.). 2007. *The Genesis of General Relativity*, 4 vols. Dordrecht: Springer:

Gutfreund, Hanoch, and Jürgen Renn. 2015. *The Road to Relativity: The History and Meaning of Einstein's "The Foundation of General Relativity" Featuring the Original Manuscript of Einstein's Masterpiece.* Princeton, NJ: Princeton University Press.

The following is the most comprehensive and up-to-date account of Einstein's scientific work:

Lehner, Christoph, and Michel Janssen (eds.). 2014. *The Cambridge Companion to Einstein.* Cambridge: Cambridge University Press.

For the reception of Einstein's theories of relativity, see the following:

Biezunski, Michel. 1991. *Einstein à Paris: le temps n'est plus. . . .* Paris: Presses Universitaires de Vincennes.

Glick, Thomas F. (ed.). 1987. *The Comparative Reception of Relativity.* Dordrecht: D. Reidel.

———. 1988. *Einstein in Spain: Relativity and the Recovery of Science.* Princeton, NJ: Princeton University Press.

Goenner, Hubert. 1992. "The Reception of the Theory of Relativity in German as Reflected by Books Published between 1908 and 1945." In J. Eisenstaedt and A. J. Kox, eds. *Studies in the History of General Relativity* (Einstein Studies vol. 3). Boston: Birkhäuser.

———. 2005. *Die Relativitäatstheorien Einsteins*, 5th ed. Munich: Beck-Verlag.

Goldberg, Stanley. 1984. *Understanding Relativity: Origin and Impact of a Scientific Revolution.* Boston: Birkhäuser.

Hu, Danian. 2005. *China and Albert Einstein: The Reception of the Physicist and His Theory in China 1917–1979.* Cambridge, MA: Harvard University Press.

Linguerri, Sandra, and Raffaella Simili (eds.). 2008. *Einstein parla italiano: Itinerary e polemiche.* Bologna: Pendragon.

Maiocchi, Roberto. 1985. *Einstein in Italia: La scienza e la filosofia italiane di fronte alla teoria della relatività.* Milan: F. Angeli.

Sanchéz-Ron, José. 1992. "The Reception of General Relativity among British Physicists and Mathematicians (1915–1930). In J. Eisenstaedt and A. J. Kox, eds. *Studies in the History of General Relativity* (Einstein Studies vol. 3). Boston: Birkhäuser, 57–88.

Wazeck, Milena. 2014. *Einstein's Opponents: The Public Controversy about the Theory of Relativity in the 1920s*. Cambridge: Cambridge University Press.

INDEX

Index entries with page numbers marked in boldface are taken from the index of Einstein's booklet.